Joy Tivy

Agricultural Ecology

D0550105

Copublished in the United States with
John Wiley & Sons, Inc., New York

Longman Scientific & Technical,
Longman Group UK Ltd,
Longman House, Burnt Mill, Harlow,
Essex CM20 2JE, England
and Associated Companies throughout the world.

Copublished in the United States with
John Wiley & Sons, Inc., 605 Third Avenue, New York, NY 10158

© Longman Group UK Limited 1990

First published 1990

British Library Cataloguing in Publication Data
Tivy, Joy
  Agricultural ecology
    1. Agriculture. Ecological aspects
    I. Title
    630′.2′745
ISBN 0-582-30163-7

Library of Congress Cataloging-in-Publication Data
Tivy, Joy.
  Agricultural ecology / Joy Tivy.
    p.   cm.
  Bibliography: p.
  Includes index.
  ISBN 0–470–21480–5 (Wiley)
    1. Agricultural ecology.   I. Title
S589.7.T58   1990
630′.2′745 – dc20

          89–12100
          CIP

Set in 10/12pt Linotron 202 Ehrhardt by
Fakenham Photosetting Ltd, Fakenham, Norfolk.
Produced by Longman Singapore Publishers (Pte) Ltd.
Printed in Singapore

Contents

Preface

Agriculture, in terms of land use, employment and food production, is still the most important of human economic activities. Like fishing and forestry, it differs from other industries in that it is a biological process: the primary products are organic and the resource base is the physical environment of land, water and air. The way in which the available resources are exploited, however, is dependent on the political system, the economic conditions and the level of scientific knowledge and technical skill prevailing. These may either inhibit a particular type of agricultural activity in an otherwise favourable physical environment or stimulate another in an area physically marginal for production. Nevertheless, the physical resource base for most types of agriculture determines the range of uses to which land can be most economically put at a particular time; and except in the controlled environments of the glass and animal houses the natural variability of the physical environment makes agriculture one of the most unpredictable of human activities.

Agriculture as a field of study is particularly wide and diverse, involving the physical, natural and social sciences. It has, in common with every other discipline, become increasingly specialized and compartmentalized. Developments in the socio-economic, agronomic, botanical, pedological and climatological aspects of the subject have all made considerable contributions to the theory and practice of agriculture. The ecological dimension has only relatively recently begun to attract serious consideration and study. However, increasing intensification of agriculture in the developed and escalating population growth in the developing areas of the world have, particularly in the last 25–30 years, focused attention on the environmental implications of the demands for increased food production. The importance of the ecological factor in agriculture is fast gaining recognition.

The ecology of agriculture is basically the study of the nature of the reciprocal interactions between the agricultural or agriculturally associated organisms and their physical habitat; and to this end the ecosystem concept is as applicable as to any other part of the ecosphere. The agro-ecosystem provides the conceptual framework within which the effects of human manipulation of crops, livestock, soil and climate on the basic ecological processes of energy flow, nutrient cycling, competition and ecoregulation can be effectively studied.

The aim of this book is to analyse some of the most important ecological characteristics and problems of agriculture with an emphasis on those factors affecting the relationships between crops/livestock and the environment and on the extent to which man has managed and modified the agro-ecosystem to his particular needs. Thus it falls into two parts. The first is systematic in approach and is concerned with the components (crops, livestock, climate, soil) and the processes of nutrient cycling and energy flows in crops and livestock of agro-ecosystems. The second part considers ecological charcteristics and the inherent or man-created problems associated with areas where climate and/or soil impose severe limitations on agricultural use or where a particular type of agriculture is dominant as in the case of paddy rice, irrigation agriculture and modern intensive farming; and concludes with a consideration of agricultural and environmental impacts.

The author lays no claim to being either comprehensive or definitive on the subject of 'agricultural ecology', the identification and conceptualization of which as a discipline in its own right is a challenge still waiting to be tackled. This book has been written in the hope that it may interest students of whatever agricultural and/or environmental persuasion and encourage them to put their specialist interests in a broader context; and that it may give food for thought to those who tend to assume that man's scientific and technical expertise is such as to minimize the significance of environmental factors in the nature and location of agricultural activities.

The author's own interest in agricultural ecology developed early in her career as a biogeographer concerned with the use and misuse of organic resources and who has become increasingly concerned by the relative neglect of the agro-ecosystem in teaching and research in biogeography. She acknowledges her debt of gratitude to those many students who have stimulated her 'environmental thinking' in this direction; and to all those friends and colleagues without whose encouragement this book might never have been written. She would like in particular to thank Professor Ian Cunningham (formerly Professor of Agriculture, University of Glasgow); Dr Joan Mitchell (Bagbee Farm, Galloway); and Drs M Kent and A Williams (Department of Geography, Plymouth Polytechnic), all of whom commented most constructively on various parts of this book. She alone is responsible for any misconception or factual errors therein.

Acknowledgements

We are grateful to the following for permission to reproduce copyright material:

Academic Press and the respective authors for figs. 5.8 from fig. 12, p. 315 (Sanchez & Salinas, 1981), 6.7 from fig. 2, p. 3 (Hudson, 1977), 9.4 from fig. 6, p. 363 (Jewell, 1980), 11.1 from table 4, p. 63 (Yoshida, 1977) and tables 6.2, 6.4, 6.5 & 7.1 from tables 7.5, 7.4, 7.7 & a table, pp. 172, 171, 174 & 109 (Spedding, 1975), 9.1 (Harris, 1980), 9.7 from table 2, p. 180 (Dyson-Hudson, 1980), 13.6 from table 1, p. 14 (Marshall, 1972); Agricultural Education Association for fig. 6.1 from fig. 1, p. 9 (Monteith, 1965); American Association for the Advancement of Science and the author, Dr. R. B. MacDonald for fig. 3.2 from fig. 1, p. 671 (MacDonald & Hall, 1980) copyright © 1980 by the AAAS; American Society of Agronomy, Inc. for table 6.1 adapted from pp. 224–226 by Singh & Stoskopf, volume 63, No. 3, May–June 1971 (Donald & Hamblin, 1976); American Society of Agronomy, Inc., Crop Science Society of America, Inc., and Soil Science Society, Inc. for fig. 1.2 from fig. 1, p. 145 (Smith & Hill, 1975) and tables 6.10 from table 2, p. 280 (Oelsligle et al., 1976) 6.11 from table 1, p. 42 (Lewis & Phillips, 1976); Edward Arnold Publishers Ltd. for table 1.1 from tables 7.14, 7.16 & 7.17, pp. 75–77 (Pimentel, David & Marcia, 1979); Blackie & Son Ltd. for table 14.1 from table 1.12, p. 14 (Slesser, 1975); Blackwell Scientific Publications and the author, R. E. White for table 5.4 from table 10.4, p. 158 (White, 1987); Blackwell Scientific Publications and the editor, W. Holmes for tables 14.2 from table 2.6, p. 14 (Williams, 1980), 14.3 & 14.4 from tables 3.7 & 3.8, p. 100 (Osbourn, 1980); Butterworth & Co. (Publishers) Ltd., for table 5.9 (Sanchez et al.), © Butterworth & Co. (Publishers) Ltd.; Butterworth Scientific Ltd. for Chapter 5, modified from Applied Geography, Vol. 7, No. 2, pp. 93–113; C.A.B. International for figs. 9.2 from fig. 3 (Norton, 1982), 14.3 from fig. 10.2 (Spedding & Diekmahns, 1972); Cambridge University Press for figs. 6.6 from fig. 4.2, p. 81 (Murata & Matsushima, 1975), 6.8 & 6.10 from figs. 1.1 & 1.2, pp. 86 & 87 (Evans, 1975), 10.7 from fig. 14.1, p. 218 (Norman et al., 1984) and tables 6.9 from a table, p. 189 (Cooper, 1975), 9.2 from table 3.1, pp. 42–43 (Caldwell, 1975), 9.8 from table 2, p. 112 (Butterworth & Lambourne, 1987), 9.9 from table 2, p. 107

(Vera *et al.*, 1987), 10.4 from a table, pp. 132–153, & 13.1 from table 21, p. 203 (Young, 1976), 10.2, 10.6, 10.7, 10.8 & 11.5 from various tables, pp. 78–159 (Norman *et al.*, 1984); Chapman & Hall Ltd. for tables 13.3 & 13.4 from a table pp. 283 & 284 (Yaron *et al.*, 1973); J. M. Dent for table 2.2 from table 4.2, pp. 70–71 (Pyke, 1970); Elsevier Applied Science Publishers Ltd. for figs. 2.1 & 2.3 from figs. 2 & 1, pp. 47 & 45 (Blaxter & Fowden, 1982); Elsevier Science Publishers (Physical Sciences & Engineering Div.) and the respective authors for figs. 5.3 & 5.6 from figs. 53 & 2, pp. 310 & 18 (Frissel, 1978), 7.6 & 7.8 from figs. 10.1 & 10.10, pp. 173 & 192 (Balch & Reid, 1976) and tables 4.12 from table 33, p. 132 (Smith, 1975), 5.8 from p. 66 (Frissel, 1978), 6.3 from table 8.3, p. 135 (Holliday, 1976); Farming Press and the author, Dr. D. B. Davies for fig. 4.5 from fig. 8a, p. 109 and tables 4.8 & 4.10 from tables 32 & 34, pp. 209 & 245 (Davies *et al.*, 1982); Food and Agriculture Organization of the United Nations for figs. 3.1a & 3.1b from figs. 3.1 & 3.2, pp. 22 & 23 (FAO, 1978), 12.2 from *World Map of Desertification*, UN Conference on Desertification, Nairobi, Kenya, 1977 (Barke & O'Hare, 1984) and tables 2.3 (FAO, 1982), 3.1 & 3.2 from tables 3.1–3.5 (FAO, 1978); W. H. Freeman & Co. for fig. 7.4 (Cox & Aitkin, 1979); the author, Professor H. F. Heady for fig. 9.6 (Heady, 1975); the Controller of Her Majesty's Stationery Office for figs. 4.2 & 13.3 from pp. 14 & 15 (MAFF, 1982), 4.3 from fig. 4, p. 68 (Eagle, 1975), 4.4 from fig. 1, p. 364 (Low, 1975), 4.6a & 4.6b from figs. 2 & 3, pp. 130 & 131 (Spoor, 1975), 4.7 from fig. 2, p. 63 (Davies, 1975), 4.8 from p. 30 (Harrod, 1975) and tables 4.2, 4.6 & 4.7 from tables 1, 7 & 3, pp. 77, 82 & 79 (Wilkinson, 1975), 4.9 from pp. 4 & 6 (MAFF, 1983), 15.1, 15.2 & 15.3 from tables 3, 6 & 23, pp. 10, 14 & 41 (MAFF, 1976), 15.4 from table 3 (DOE, 1984); Institute of Biology for table 15.6 from table 1, p. 2 (Strickland, 1960); International Rice Research Institute for fig. 11.2 (Crasswell & Vlek, 1979), fig. 11.3 and table 11.3 from table 2, p. 49 (Morrmann & van Breeman, 1978); International Thomson Publishing Services Ltd. and Sydney University Press for fig. 3.4 from fig. 11.5 (Jeans, 1977); Kluwer Academic Publishers for fig. 7.7 from fig. 4, p. 289 (Klingauf, 1981); Longman Group UK Ltd. for figs. 4.9 from fig. 7.5, p. 99 (Simpson, 1980), 7.3 from fig. 1.2, p. 16 (Mason, 1984), 9.5 & 13.2 from figs. 10.4 & 13.1, pp. 145 & 192 (Heathcote, 1983) and tables 11.2 from *Tropical Crops: Dicotyledons* by Purseglove, 1968 (Wrigley, 1981), 11.4 from table 20.9, p. 489 (Grist, 1975); Longman Group UK Ltd. for fig. 1.1 from fig. 5.3, p. 86 (Tivy & O'Hare, 1981) published by Oliver & Boyd; Lunds Universitet for fig. 12.3 (Rapp & Helleden, 1979); Macmillan Publishers Ltd. for figs. 5.4 & 5.5 from figs. 21.1 & 17.2, pp. 531 & 439 (Buckman & Brady, 1969); Methuen & Co. for table 13.2 from p. 115 (Hills, 1966); Ministry of Agriculture Fisheries & Food for table 5.6 from *ADAS Booklet No. 2081* © Crown Copyright 1985 (Archer, 1985); The Open University for figs. 9.3 from p. 59, 14.2 from figs. 23 & 25, pp. 47 & 52, 14.4 from figs. 26 & 27, pp. 52–53 (Morris, 1977) © 1977 The Open University; Oxford University Press for fig. 10.6 from fig. 3.15, p. 80 (Kowal & Kassam, 1978) and tables 9.5 from p. 379 (Schmit-Neilson, 1956), 9.6 & 10.9 from tables 9.1 & 8.1, pp. 298 & 239 (Ruthenberg, 1976); Pergamon Press PLC for table 4.11 from p. 186 (Cannell & Finney, 1973) © 1973 Pergamon Press PLC; George Phillip & Son Ltd. for table 10.3 from table 2, p. 233 (Young, 1974); Royal Agricultural Society of England for fig. 7.5 from fig. 20.1, p. 366 (Greenhalgh, 1977); Royal Netherlands Society for Agricultural Science for figs. 6.2 (Alberda, 1962), 6.4 from fig. 19.5, p. 336 (Sibma, 1977); The Royal Society and the respective authors for figs. 13.1 & 13.4 from figs. 3 & 2, pp. 266 & 263 (Stanhill, 1986) and tables 4.4 from table 1, p. 196 (Greenland, 1977), 5.3 & 5.5 from tables 4 & 1, pp. 215 & 97 (Russell, 1977), 5.10 & 5.11 from tables 5/6 & 10, pp. 235 & 239 (Cooke, 1977), 6.6 from table 2, p. 316 (Hood, 1982), 6.8 from table 6, p. 208 (Reece, 1985), 7.3, 7.5 & 7.6 from tables 5, 6, 1, 3 & 7, pp. 122–126 (Holmes, 1977); Scientific American, Inc. for fig. 6.9 from p. 186 (Christie in Jennings, 1976) copyright © 1976 by Scientific American, Inc.; the author, Dr. R. Smith for table 8.1 from p. 47 (Smith & Atkinson, 1975); Soil Survey and Land Research Centre and the Macaulay Land Use Research Institute for tables 4.5 from p. 135 (Brown, 1974), 8.4 & 8.6 from pp. 3 & 4 (Bibby & Mackney, 1969), 8.5 from p. 3 (Bibby, 1982); The Unesco Press and the author, Dr. V. J. Valli for table 3.6 from table 1, p. 348 (Valli, 1968) © 1968 The Unesco Press; United States Department of Agriculture (Soil Conservation Service) for table 8.3 (Klingebiel & Montgomery, 1961); John Wiley & Sons Inc. for figs. 5.2 from fig. 5.1, p. 174 (Sanchez, 1976), 10.1, 10.2, 10.3, 10.4 & 10.5 from figs. 1.1, 1.2, 8.2, 8.6 & 10.1, pp. 3, 106, 113 & 191 (Nienwolt, 1975) and tables 10.5a & 10.5b from tables 8.23 a & b and 8.24 a & b, pp. 514 & 515 (Aina *et al.*, 1979), 11.1 (de Datta, 1981), 15.5 from table 11.1, p. 111 (Dix, 1981).

Whilst every effort has been made to trace the owners of copyright material, in a few cases this has proved impossible and we take this opportunity to offer our apologies to any copyright holders whose rights we may have unwittingly infringed.

1

The agro-ecosystem

Agriculture, in its widest sense, can be defined as the cultivation and/or the production of crop plants or livestock products. It is generally synonymous with 'farming' – the field or field-dependent production of food, fodder and industrial organic materials. In common with the other traditional primary industries (forestry, fishing) the primary resource base for agriculture is the physical environment. The basic process involved is that of photosynthesis – a process which has not yet been replicated outside the living chloroplast-containing cells of green plants. The crop plant is, then, the fundamental 'production unit' in agriculture because of its ability, given sufficient light energy, to manufacture complex organic materials (carbohydrates, proteins, fats) from simple inorganic elements such as carbon dioxide, water and mineral nutrients supplied by the atmosphere and the soil. Plant productivity (or net primary biological productivity) is the rate at which the plant accumulates organic matter surplus to its own energy requirements; and provided there is neither a deficiency of water nor a lack of nutrients this will be a function of the intensity and duration of insolation received. The minimum, optimum and maximum environmental thresholds for the growth and development of different species or varieties determine the area within which a particular crop plant can be successfully cultivated either to provide sufficient food to meet immediate requirements or to give a satisfactory economic return from the yield obtained.

The cropping, whether of plant or animal products, requires a degree of management of the available environmental and organic resources. The primary management practice is cultivation of the land or, more precisely, the soil. Cultivation involves:

1. *Selection* of that crop plant best adapted to ensure a high probability of a satisfactory yield of the required product.
2. *Propagation*: the preparation of the soil by some form of *tillage* in order to ensure suitable conditions for planting or sowing and for 'feeding' the crop.
3. *Protection* from competition for the primary resources by weeds and from the

direct or indirect reduction in yield potential by animal pests and pathogenic organisms.

Cultivation involves management of both the crop and the environment (or habitat) in which it is grown. On the one hand, crop selection and breeding aims to produce a plant that will give the highest yield and/or quality under a given set of environmental conditions. On the other hand, soil tillage, soil-water management, weeding and pest control aim to minimize biophysical constraints and to produce a habitat which will allow the crop to realize its productive potential. In other words the farmer works within the limits of his inherited or acquired cultural and technical abilities to achieve the 'best fit' between the crops he chooses to grow and the physical habitat. In doing so he 'creates' a particular type of ecosystem – the agricultural or, as now designated, the *agro-ecosystem* in which the farmer is an essential ecological variable in influencing or determining the composition, the functioning and the stability of the system.

The *agro-ecosystem* differs from other 'wild' unmanaged climax ecosystems, in a similar physical environment, in being simpler, with less diversity of plant and animal species and with a less complex structure (i.e. spatial organization of its organic components). Not only are the number of species and associated life-forms fewer but species or varieties of crop plants tend to have less genetic diversity than their wild ancestors or near relatives. Because of this simplification, incoming solar energy is channelled along fewer and shorter routes or food-chains than in the wild ecosystem. Cultivation, by reducing competition, channels a high proportion of the available light energy into the crop plants and from there directly, or indirectly, by way of domestic livestock products, to man. The result is a reduction in the complexity of the 'food-web' and in the number of trophic levels to two or, at most, three. Also, in the agro-ecosystem the biomass of large herbivores (such as cattle, sheep, goats etc.) is considerably greater than that of ecologically equivalent animals normally supported by the unmanaged terrestrial ecosystem. As a result a much smaller proportion of the energy of the plant biomass is passed along the 'detrital or decomposing route' in the soil. In contrast, a higher proportion is consumed by the domestic livestock or is exported from the system as a plant or animal crop. Consequently the 'energy pool' in the soil in the form of dead and decaying organic matter and humus is, in general, much less than that in unmanaged ecosystems in similar environments.

In addition, the more highly concentrated energy flow in the agro-ecosystem is accompanied by a disruption of the natural nutrient cycle. In the latter, most of the nutrients taken up from the environment by plants are returned and recycled by the decomposition of dead vegetation and animals and the excrement of the living animals. In the agro-ecosystem there is, inevitably, a continual slow or rapid loss of nutrients from the system either in the crop or animal harvest or as a result of the increased rate of organic decomposition and of leaching of nutrients consequent upon soil tillage. Also, in short-life crops or livestock the rate of nutrient cycling tends to be speeded up and the living nutrient pool is more variable in volume and is dependent more on the animal than the plant biomass, in contrast to the unmanaged system. Unless nutrient loss through export is made good by man in the form of organic or inorganic fertilizers, the agro-ecosystem will inevitably tend to run down, the biomass will decrease and its productivity will fall. However, with increasing knowledge and techniques, man has developed ways of either maintaining or increasing the productivity of agro-ecosystems by breeding high-yielding crops and animals and increasing the input of nutrients, while at the same time attempting to reduce the loss of yield as a result of competition from weeds and pests.

Unmodified or 'wild' ecosystems are frequently attributed with a higher degree of stability than agro-ecosystems because their greater species diversity and trophic complexity is thought to allow them to attain homeostasis, i.e. to achieve and maintain a steady-state in the face of normal environmental variability. In contrast, agro-ecosystems, lacking this inbuilt regulatory mechanism, are very susceptible to disruption or even to destruction by extreme environmental conditions – drought, frost, disease etc. It has, however, also been postulated that such a comparison is invalid because the stability of an agro-ecosystem is not dependent so much on 'natural' regulatory mechanisms but on how successful man is in managing the system and, thereby, insuring against the normal and predicting the probability of abnormal environmental variations.

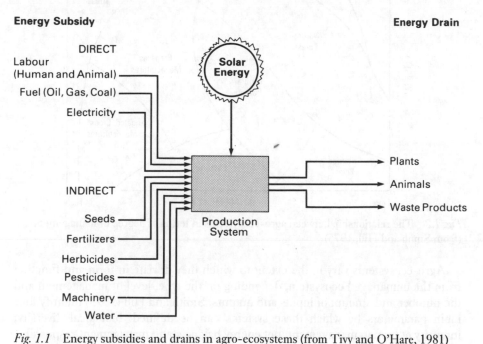

Fig. 1.1 Energy subsidies and drains in agro-ecosystems (from Tivy and O'Hare, 1981)

One of the most striking and ecologically significant contrasts between the unmodified ecosystems and the agro-ecosystem is that the latter is a very much more *open system*, with a greater number and a larger volume (or energy equivalents) of inputs (gains) and outputs (losses), than the former. In terms of inputs agro-ecosystems are 'subsidized' by the addition of either the *direct* or *indirect energy* which is needed for cultivation and for increased yields (Fig. 1.1). In addition, the primary or secondary production from one agro-ecosystem may become an input for another. Feedstuffs grown on one farm are consumed on another; or livestock reared on one farm are fattened on another. Similarly the number and volume of outputs from agro-ecosystems exceed those from wild ecosystems. As well as the 'export' of plant or animal 'crops' the former systems produce much more waste: first, because the harvestable crop normally represents only a small but variable proportion of the total plant; and secondly, because densities of livestock are so much higher than those of herbivores in unmanaged ecosystems, the amount of animal excreta produced is correspondingly larger. Depending on the type of farm system, some of the organic waste is recycled, i.e. returned to the land as manure or organic fertilizers; however, much, together with surplus inorganic fertilizers and chemical pesticides, can 'leak' from the system in which it has been produced or applied, and become an input, often a polluting input, into other wild or managed ecosystems.

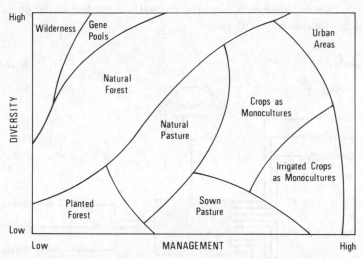

Fig. 1.2   The relationship between agro-ecosystems in terms of degree of management (from Smith and Hill, 1975)

Agro-ecosystems vary in the extent to which they deviate in form and function from the unmanaged ecosystem, depending on the type, level of management and the number and amount of inputs and outputs. Smith and Hill (1975) identify four main parameters by which these systems can be defined: biological diversity; intensity of human management; net energy balance; and management responsibility; and they stress that there is a continuum between the wild or unmanaged and the most intensively managed ecosystem (Fig. 1.2). At one end of the spectrum, where the level of management is low and input other than human labour is negligible, the existing semi-natural ecosystem may be directly exploited for livestock production, as in open-range grazing, with little management of either its organic components or the physical environment. In this case, inputs and outputs are small and agricultural productivity is closely related to that of the uncultivated vegetation resource. Similarly in one of the simplest forms of cropping, that of shifting agriculture in the humid tropical rainforest environment, management is limited and the particular agro-ecosystem is closely adapted to the existing ecological conditions. It has been suggested that this agro-ecosystem, in the diversity of the species and life-forms cultivated, replicates in miniature that of the natural forest (Harris, 1976; Ruthenberg, 1980). At the other end of the spectrum are the agro-ecosystems in which there is a very high level of management involving often drastic modification of the environment, crops and livestock. The high number and volume of capital, rather than labour, inputs is reflected in high productivity. Indeed it has been suggested that, in this type of agro-ecosystem, man's technical expertise is such that the physical environment is no longer a significant variable either determining or influencing the type of agro-ecosystem. This is certainly valid for those crops and livestock produced in the controlled environment of the glasshouse, the hen battery and the piggery. Nevertheless, the cost of controlling the environment 'indoors' may, to a greater or lesser extent, be influenced by light and heat gains from and losses to the surrounding environment.

Table 1.1. illustrates varying intensities of rice production in terms of the type and amount of direct and indirect energy subsidies. In Borneo, methods of cultivation are simple and the most important input is that of human energy. Yields are low and will in such systems be dependent on the inherent fertility of the soil and how successfully the natural nutrient cycle is maintained. In Japan, although rice cultivation is more labour-intensive than in Borneo, a higher proportion of the direct energy subsidy is accounted for by mechanization and indirect energy subsidies are over twice those from direct energy. Fuel inputs are one to two and a half times greater in Japan than in Borneo. The level of all inputs is highest in California, where machinery has all but displaced human labour and where all other inputs, with the exception of phosphorus and calcium, exceed those in Japan. Energy output of yields is three times higher than in Borneo. The productivity of all three areas is a function of the level of inputs, that is, of the intensity of agriculture. However, there is an inverse relationship between productivity and energy efficiency, the energy efficiency of rice production in Borneo being nearly five times greater than that in California. With increasing intensity, labour efficiency in terms of output per unit input of kinetic energy or per man–hour of work increases as does the cost of production.

Finally, the scale of the agro-ecosystem can vary from small, involving intensive land use, to large, dependent on an extensive land area. The scale is also dependent on the degree of openness of the agro-ecosystem. Some, as in the case of subsistence agricultural economies, are more self-contained than others. They have little or no energy subsidization other than that provided by human labour and simple hand-implements. Others, as exemplified by the most intensive, not only are very open and highly subsidized but the scale of their operation may be global in terms of the distance from which energy subsidies are derived and to which outputs (products and waste) are exported.

The type of agro-ecosystem is determined partly by the environmental potential and partly by the prevailing social, economic and political conditions which influence the demand for and profitability of the crops and livestock that can be

Table 1.1 Energy inputs and outputs in rice production

	Energy (000s kcal ha^{-1})		
	Borneo	Japan	California
Inputs			
Direct energy:			
Labour	0.626	0.804	0.008
Axe and hoe	0.016	–	–
Machinery	–	0.189	0.360
Diesel	–	–	3.264
Petrol	–	0.910	0.657
Gas	–	–	0.354
Indirect energy:			
N	–	2.088	4.116
P	–	0.225	0.201
Seeds	0.392	0.813	1.140
Irrigation	–	0.910	1.299
Insecticides	–	0.348	0.191
Herbicides	–	0.699	1.119
Drying	–	–	1.217
Electricity	–	0.007	0.380
Transport	–	0.051	0.121
Outputs			
Rice yield	7.318	17.598	22.3698
(Protein yield)	(141 kg)	(364 kg)	(462 kg)
Energy efficiency	7.08	2.45	1.55

(from Pimentel, David and Marcia, 1979)

produced in a given area. The optimum location for a particular farming system will be that environment in which the maximum return for inputs can be achieved. While its potential distribution is determined by environmental conditions, its actual limits may be constrained by socio-economic conditions or extended by techniques which minimize or overcome physical limiting factors or by economic protection. However, as Loomis and Gerakis (1975) note, farmers tend to be conservative and their choice of system is influenced by the need to achieve as high a probability of success as possible in what is environmentally a high-risk undertaking.

Crops

The number of plant and animal species used by man is minute in comparison to the total available. For example only about 3000 of the 350 000 species of flowering plants have assumed any economic significance. Further, the number used has decreased dramatically with man's transition from a hunter-gatherer to a pastoralist or sedentary farmer and from a subsistence to commercial farmer. Initially the human diet was very much more varied than it is today. Harlan (1976b) has suggested that formerly several thousand plants and hundreds of animal species were eaten. An analysis of the contents of the stomach of the Iron-Age 'Tollund Man' (Helbaek, 1950) revealed that his last meal had been composed of some seventeen different plants of which only two, oats and barley, were cultigens; the remainder were the weeds of cultivation. Today twelve crop plants and three types of livestock provide the bulk of the world's food.

The development of agriculture is a relatively recent event in human evolutionary history. It has been dependent upon the selection and propagation – the *domestication* – of species or varieties of plants and animals with characteristics which make them particularly desirable and/or useful to man. This not only has altered the relative composition of the world's plant and animal population but, over at least 10 per cent of the earth's surface, has resulted in the replacement of the pre-existing wild by a cultivated vegetation cover and in the drastic modification of much of the remaining area by the grazing of domestic livestock. The agro-ecosystems that have resulted are simpler in composition and structure than the wild or little-modified ones. They are dominated, to a greater or lesser extent, by three main types of organisms:

1. Cultivated (crop) plants or cultigens.
2. Domestic animals.
3. Associated weeds, pests and pathogens.

The domesticated plants and animals are those whose propagation and breeding are controlled, to a greater or lesser degree, by man. Derived originally from existing

7

wild organisms, selection and breeding has tended, on the one hand, to eliminate those features unattractive to man but which may have had a particular use or survival value in the wild progenitors and, on the other, to develop characteristics which are of particular agronomic value (Table 2.1). In the most highly domesticated species the 'symbiotic' relationship with man is such that few would survive in the wild without his protection.

Table 2.1 Twenty most important crops according to area occupied

Common name	Proper name	Agronomically significant part	Main uses
1. Wheat*	(*Triticum aestivum*)	cereal	food, fodder
2. *Rice* (paddy)	(*Oryza sativa*)	cereal	food
3. *Maize*	(*Zea mays*)	cereal	fodder
4. Barley	(*Hordeum sativa*)	cereal	fodder, food, malt
5. *Pearl millet*	(*Panicum miliceum*)	cereal	food, fodder
6. *Sorghum*	(*Sorghum* spp.)	cereal	food, fodder
7. *Soya bean*	(*Glycine max*)	seed-legume	fodder, food, oil
8. *Cotton*	(*Gossypium* spp.)	seed	fibre, fodder, oil
9. Oats	(*Avena sativa*)	cereal	fodder, food
10. *Field bean*	(*Phaseolus* spp.)	seed-legume	fodder, food
11. *White potato*	(*Solanum tuberosum*)	tuber	food
12. Groundnut	(*Arachis hypogea*)	seed-legume	oil, fodder, food
13. Rye	(*Secale cereale*)	cereal	food, fodder
14. *Sweet potato*	(*Ipomoea batatus*)	root	food
15. *Sugar cane*	(*Saccharum officinarum*)	stem	food, fodder
16. *Cassava*	(*Manihot* spp.)	tuber	food
17. Pea	(*Pisum sativa*)	seed-legume	food, fodder
18. Chick pea	(*Cicer arietinum*)	seed-legume	food, fodder
19. Grape		fruit	food, beverage
20. Oil-seed rape	(*Brassica napus* forma *oleifera*)	seed	oil, fodder

* Crops shown in italics have the widest regional distribution. Percentage of total cropland area occupied by cereals 71.0; oil-seeds 7.2; roots and tubers 5.0; pulses 4.9; fibres 4.7; fruit and vegetables 3.7; sugar 1.5; beverages 1.0; rubber 0.4; tobacco 0.4.

THE CROP PLANT

The value of a crop is dependent on its ability to produce a larger amount of useful material than its wild progenitor and on its greater ease of cultivation. One of the most striking and consistent differences between the crop and the wild plant is the 'gigantism' of the former (Schwanitz, 1966). This is expressed particularly in those organs selected for harvesting and in the proportion they comprise of the whole plant. In the cereals, for instance, a higher proportion of material assimilated during photosynthesis goes into the grain than into the leaves, stem or root. Conversely, in the potato the main storage organ is the tuber. Another important and closely associated characteristic which distinguishes the cultivated crop from its wild progenitor is a reduction in natural seed dissemination as a result of selection for tough non-shattering flower-stems in the cereals or indehiscent pods and fruit in peas, beans, peppers etc. Concomitant with reduced dissemination has been a very marked increase in seed size; for example the seeds of the common field bean (*Phaseolus vulgaris*) are five to eight times larger than those of the wild species. The composition, as well as the form, of organs selected for harvest has also changed with domestication. Toxic and bitter substances, as in the yam and lupin, have been reduced or eliminated. The proportionate content of substances of value to man, such as proteins, oil, sugar, drugs or aromatics, has been considerably increased.

Crop plants are often classified on the basis of their particular use or uses. These, however, are highly variable and susceptible to change. A more rational grouping, of universal applicability, is that on the basis of the harvested organ, as indicated below and in Table 2.2:

Cereal (grains)	– seed
Seed-legumes	– seed
Roots and tubers	– swollen roots and stems
Vegetables ('green crops')	– leaves and stem
Fruit	– seed coat or capsule

The *seed* of cereal and of seed-legume crops is the most important primary source of carbohydrates, fats, proteins, vitamins and minerals essential for human and livestock nutrition. With a low (less than 15 per cent) water content and a relatively long dormancy period, the cereal seed provides a highly concentrated source of food that in addition is both easily stored and transported (see Table 2.2).

Table 2.2 Composition (per 100 g) of the more important cereal grains in comparison with that of a tuber, seed-legume and green vegetable

	Energy (kcal)	Water (%)	Protein (g)	Fat (g)
Wheat	340	8.2	12.0	2.0
Rice (brown)	310	12.3	8.0	2.0
Maize (yellow)	352	12.0	10.0	4.0
Sorghum	348	–	10.0	4.0
Oats	317	–	10.0	5.0
Rye	338	–	11.0	2.0
Potatoes (white)	83	77.8	2.0	0.1
Peas (green; cooked)	70	81.7	4.9	0.4
Lettuce	15	94.8	1.2	0.2

(from Pyke, 1970)

Roots and *tubers* are, in contrast, bulky organs with a much greater water content (over 75 per cent), which limits storage and makes the harvesting effort per unit material produced high. The majority (with the exception of the potato) are biennials which are cropped at the end of the first season's growth when the storage organs have reached full size.

Vegetables comprise all those crops whose value is related to their fresh leaves and young stems. A high cellulose and fibre content limits their use for food, though they are an important source of vitamins and minerals, and of the social drugs such as nicotine, tannin and caffeine. Large bulk and high water content, however, make their storage dependent on drying, ensilage or quick-freezing.

The term *fruit*, in the agronomic sense, is applied either to the fleshy material containing the seeds (i.e. berries, and 'pomes', such as apples and pears) or to single-seeded drupes (such as coconuts, cherries and plums). Some, such as the banana, may have a high carbohydrate content. In most fruits the economic value lies in their attractive taste as much as in a high vitamin C content. As with vegetables their water content makes storage dependent on some form of 'preservation'.

Ease of cultivation is closely dependent on the *growth form* (or habit) and the *life-cycle* of the particular crop plant. The majority of crops are propagated by seed. Good seed production combined with rapid and uniform germination are prerequisites of the successful crop plant. However, the recent rapid development of hybrid varie- ties, particularly of cereals, has made propagation increasingly dependent on the specialized cultivation of seed-producing crops. Vegetative reproduction is most common among the roots and tubers such as the white potato, and those such as the sweet potato, cassava and yam, food staples of the humid Tropics, where environ- mental conditions are not so suitable for grain production. Vegetative reproduction is also important for some fruit-producing plants such as banana, strawberry, pineapple and sugar cane.

The majority of crops, and particularly the grains, are annuals. Others are biennials which, as in the case of many root crops, are usually cultivated as annuals and are only allowed to 'bolt' (or run to seed) for the purposes of propagation. Many normally perennial plants are cultivated on a short-life basis or, in some cases such as cotton, even as annuals. Harvesting is facilitated by determinate flowering, an upright form, tough non-shattering stems and indehiscent seeds. It is therefore not surprising that the cereals – the most highly domesticated plants – have all the attributes needed by a good crop plant and, what is more, have been among the most amenable to mechanized harvesting. The most difficult have been the root crops, fruits and fresh leaves, which combine easily damaged material with indeterminate growth and maturity and a high level of seed dehiscence.

CEREALS

The most important crops (Table 2.3) in terms of value, production and area occupied are, according to Harlan (1976b), those food crops which give a high return of storable produce per unit effort; have a 'safe' (i.e. reasonably consistent) biological performance; and satisfy the basic nutritional needs of humans and

Table 2.3 Major food crops

Crop	World production (tonne × 10^{-6})	Percentage contribution to world production
Wheat	417	15.7
Rice	345	13.0
Maize	334	12.5
Potatoes	228	10.8
Barley	190	7.1
Sweet potatoes	136	5.1
Cassava	105	3.9
Soya beans	62	2.3
Sorghum	52	2.0
Seed-legume	52	1.9
Millet	52	1.9
Oats	49	1.8
Tomatoes	41	1.5
Rye	28	1.0

(from Food and Agriculture Organization Production Year Book 1982)

livestock. Among these, the *cereals* (from Ceres, the Goddess of Grain) or *grain crops* are pre-eminent (see Fig. 2.1). They comprise the *small grains* (wheat, rice, barley, rye, oats) and the *large* or *coarse grains* (maize, sorghum, millet). Collectively these provide over 50 per cent of the world's energy and protein needs (over 75 per cent if grain fed to livestock is included) and occupy two-thirds of the cultivated land. The earliest domesticated crop plants, the cereals have long been the staple food of humans. Grains dominate the bulk and value of the world's *food trade*; and the world's *food reserves* (i.e. food in store) today are expressed annually in terms of the amount of grain left over after trading has finished. The importance of the cereals is undoubtedly related to their relatively high nutritive value (protein content *c.* 10–12 per cent); their ease of cultivation, storage and transport; their wide range of adaptation to environmental conditions; and their early maturity. The three principal and oldest grains – rice, wheat and maize – each comprise a vast complex of *cultivars* and strains, the result of selection and breeding with a history almost as long as that of man himself.

Wheat and rye are the 'bread' cereals. Both contain a slightly higher amount of protein than the other grains. Wheat contains a higher proportion of the protein – *gluten* – which allows dough to rise, and as a result the flour produces a lighter, more porous bread with a higher energy value than does rye; it is also more widely used. Rye flour, however, has a high nutritional status associated with its lysine-rich protein content, although its calorific value is lower than that of wheat. The 'black bread' of *rye* has been, and still is, equated with the poor man's diet, particularly in Eastern Europe, though it has been introduced by immigrants as a speciality 'health' food, into more affluent societies in North America. Rye has the advantage of being more frost-hardy and less exacting in its soil requirements than any of the other

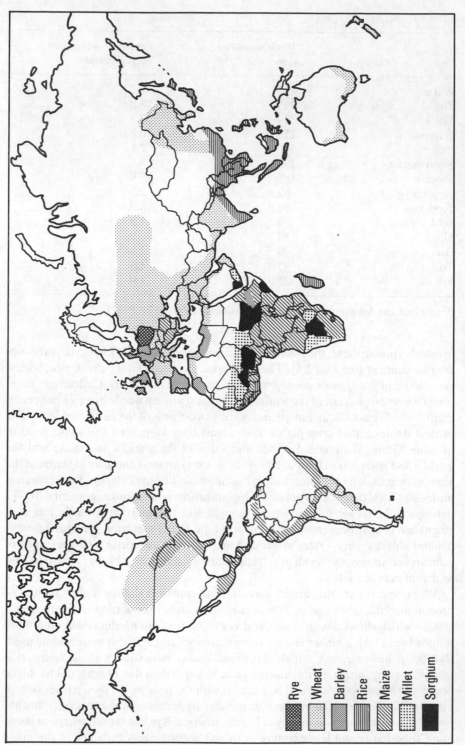

Fig. 2.1 Predominant cereal crop, by area occupied, for each country in 1978 (redrawn from Blaxter and Fowden, 1982)

Rye
Wheat
Barley
Rice
Maize
Millet
Sorghum

cereals. Although declining in acreage, it is still grown extensively in Eastern Europe and, as well as providing feedstuff, it is an important cereal for the distillation of alcohol. A hybrid, *Tricale*, has been recently produced from a cross between wheat and rye. It combines the protein content of wheat with the lysine quality of rye. However, because of its deficiency in gluten, only a few breeding lines have so far produced flour suitable for bread. It is nevertheless hoped that cultivars superior in yield to rye but more hardy than wheat may eventually be successfully established.

Wheat and *rice* are the most important food grains. The former has a higher protein content, though the quality depends on the variety and the environmental conditions under which it is grown. With a wide range of adaptability to climatic and soil conditions it is more extensively cultivated than any other food crop. Deeply rooting, it grows well on heavy soils with a high nutrient status. Some 75–80 per cent of the world's wheat crop is comprised of the higher yielding *winter-sown* varieties, which need low temperatures to initiate successful spring germination and ensure rapid maturation. *Spring-sown* varieties require a longer growing season, which, together with lower potential yield, restricts their range. In areas of low rainfall and high summer temperature, relatively low yields are compensated by the high protein values of the *hard wheats*. These give a gritty flour, admirably suited to bread-making. In contrast, the *soft wheats*, better adapted to humid cool-temperate climates, produce a more general-purpose flour used in pastries, biscuits, breakfast cereals and, importantly, for livestock feedstuff. In the humid Tropics wheat cultivation is limited to areas of high altitude. Susceptibility to fungal attack and problems of harvesting and storage preclude it from areas where high temperatures are associated with high atmospheric humidity.

Although *rice* does not occupy such a large area as wheat, it is nevertheless the staple food of approximately half of the world's population. Over 95 per cent is grown in the humid tropical and subtropical coastal lowlands of the Far East, where dependence on rice is greater than on any other single food crop. Like wheat it has a long history of cultivation and is today represented by over fourteen thousand varieties adapted to local environmental conditions and traditional types of farming. Two main groups are recognized – *indica* (native to South Asia), and *japonica* (native to Japan, North China and Korea), whose cultivars are adapted to the longer days and shorter growing season of these areas. Rice is unique among the cereals in its ability to grow and germinate successfully partially submerged in water. It needs high temperatures (11–12 °C minimum) for germination. However, selection of early maturing varieties allows it to be grown as far north as Hokkaido (Japan), while irrigation has extended its range into low-rainfall or formerly arid areas in many parts of the world.

The two remaining temperate cereals, *barley* and *oats*, are used predominantly as feed crops. Barley is distinguished by its ability to mature in a shorter time than any other cereal. While its climatic requirements are similar to those for wheat, it can be successfully cultivated at higher latitudes or in more arid areas than can wheat. Protein relationships in response to climate are also similar to those in wheat and are reflected in the different requirements for *feed* and *malting* barley (an important secondary use). The high starch to protein ratio of barley grown in humid conditions is more suitable for the production of malt than the flinty grain higher in protein, produced where the growing season and yields are curtailed by hot dry weather. Though not so demanding of nutrients as wheat, barley is less tolerant of poor drainage and high acidity. It can, however, tolerate a moderately high concentration of soluble and alkaline salts.

Oats (both grain and straw) are used mainly for feed. Although the nutritive value, particularly in fat (8 per cent) and protein (16 per cent), is high, the yield potential is low. The grain hull is more fibrous and silkier than that of the other cereals. Oats are, characteristically, a crop of humid cool-temperate regions – but are less hardy and less demanding of soil than either barley or wheat.

The three major grains – maize (Indian corn), sorghum and millet – are all crops of tropical origin. *Maize*, which originated as a food crop in Central America, together with wheat and rice was among the earliest food crops to be domesticated. It possesses a similarly high genetic diversity and wide range of environmental adaptation. It is the heaviest yielding cereal. The most important groups commercially today are:

Pod corn (*Zea mays tunicata*)
Flour corn (*Z. m. amylacea*)
Popcorn (*Z. m. everta*)
Flint corn (*Z. m. indurata*)
Dent corn (*Z. m. indentata*)
Sweetcorn (*Z. m. saccharata*)

High-yielding varieties of hybrid corn were developed in the 1920s and 1930s in the USA. Nearly all the corn grown today is of hybrid origin and the production of hybrid strains, adapted to growing seasons curtailed either by drought or by low temperatures, has witnessed a rapid extension of the range of corn into cooler and drier areas than before. It is now grown over a very wide range of latitude (from 58°N to 40°S) and of altitude in the Tropics (from sea level to *c.* 4,500 m). High summer daytime temperatures (*c.* 24 °C), combined with warm nights, are needed for high performance. One of the most serious limiting factors is insufficient moisture during the critical period of flowering and fertilization (i.e. silking).

In contrast to the other two major food crops, maize has become the universal feed grain *par excellence*. While a deficiency in the vital amino acid, *lysine*, seriously reduces its food value, a high fat plus protein content makes it nutritionally the most valuable of the 'concentrated' animal feedstuffs. Further, in areas marginal for grain ripening, but where vegetative growth is nevertheless rapid, corn maintains its value as a green forage crop. The grain is also a source of many other subsidiary products including breakfast cereals, cornflour, vegetable oil, and sugars which can be used in the production of alcohol.

Sorghum and *millet* are the food grains of dry tropical/subtropical climates. Sorghum is the higher yielding of the two. In addition, it is particularly drought-resistant; and is probably the only field crop approximating to a true xerophyte. Four groups are recognized on the basis of seed characteristics and main use:

1. Grain sorghum: food and feed.
2. Sweet sorghum ('sorgo'): forage, syrup, alcohol.

3. Grass sorghum: of which Sudan grass is the most important hay/pasture species.
4. Broom corn.

In addition, *kaoliang* comprises a group of hardy sorghums which are drought-resistant but can be grown under lower temperatures than the other varieties. In contrast to maize, *sorghum* is a food grain of African origin which has, particularly since the Second World War, increased in value as a fodder crop in the semi-arid areas of the South-West USA. As a result breeding to suit the needs of intensive mechanized agriculture has probably taken place more rapidly in the sorghums than in any other field crop. *Millet* plays a similar role as an important food grain in the drier Tropics, and as a feed crop in other areas. Although millet is grown less extensively than formerly, it is still an important food crop in both Africa and Asia, in areas too dry or with soils too poor to grow sorghum satisfactorily.

The importance of all the cereals is related to their ability to supply most of the human nutritional requirements for bodily growth, maintenance and health. However, even the most important staple, wheat, is slightly less than nutritionally complete in terms of the amount, balance and, particularly, *quality* of its protein make-up. Too little, combined with poor quality, protein in the human diet is the major factor which gives rise to *kwashiorkor* or the bodily conditions resulting from malnutrition and, particularly, protein deficiency. Depending on age, nine to ten different amino acids (the protein 'building blocks') are essential for health. Plant protein is usually deficient in one or more of these as well as lacking vitamin B_{12}. Lysine and methionine levels are generally too low in all grains, while maize is also deficient in tryptophan.

LEGUMES

In contrast to the cereals, the legumes are rich in lysine but low in the sulphur-containing cysteine and methionine, while the protein in leafy greens is well balanced in amino acids, except for methionine. If one or more amino acid is deficient, the utilization of all the others present will be reduced in the same proportion, because 'surplus' acids cannot be stored and are broken down for energy. A nutritionally satisfactory vegetarian diet, therefore, can only be achieved by a combination of foods which results in *protein complementarity* (Scrimshaw and Young, 1976), i.e. when protein sources are mixed in such a way that they cancel out each other's amino-acid deficiencies (see Fig. 2.2a, b). While the legumes, in comparison to the cereals, are lacking in adequate amounts of a number of essential amino acids, they have sufficient lysine to make good this most serious deficiency in the cereals, combined with a high ratio of protein to carbohydrate in their seeds. In those parts of the world where people are dependent on largely vegetarian diets, the *legumes* have long played the role of protein supplement. For this reason they have often been called the 'poor man's meat'.

After the cereals, the *seed-legumes* – soy, peanuts (groundnuts), peas, beans – are the most widely cultivated crops in both temperate and tropical latitudes. They have three agronomic features in common. First, they all produce seeds in a fruit 'pod',

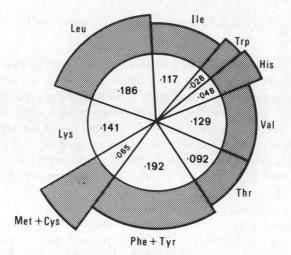

Fig. 2.2a Proportion of essential amino acids present in wheat flour. Number + amount of amino acid that can contribute to protein synthesis. All lysine is used; the surplus amino acids (*shaded portion*) cannot be stored and are oxidized. Leu = leucine; Ile = isoleucine; Val = valine; Phe = phenylalanine; Tyr = tyrosine; Thr = threonine; Lys = lysine; Met = methionine; Cys = cysteine; Trp = tryptophan; His = histidine (redrawn from Scrimshaw and Young, 1976)

i.e. the *legume* which has given its name to the group. Secondly, they produce hard, and in the most important agricultural species relatively large, seeds (called 'pulses') in which protein rather than carbohydrate is the main storage compound. Thirdly, they nearly all can fix atmospheric nitrogen (via the symbiotic bacterial group, *Rhizobium*, in their roots) at an average rate of between 150 and 250 kg ha yr^{-1} and up to 600 kg ha^{-1} yr^{-1} in some instances. Many leguminous species also have a high fat content and are among the most important sources of vegetable oil. In addition, they provide a particularly nutritious animal feedstuff (green or dry forage) and are also extensively used as 'cover crops' and soil improvers.

The three most important food-producing seed-legumes are the soya bean, the common field bean and the groundnut (or peanut). *Soy* is the most widely cultivated and most versatile of the three. It is native to Japan, China and India, where it is primarily used for food. In the 1930s it was introduced into the USA as a soil-conserving cover crop, and it later became an important multipurpose field crop there. It was easily adapted to both the warm humid summer climate and the intensively mechanized corn-based agriculture of the Central USA where it is an important feed grain. Since the Second World War, soy has emerged as an important 'industrial' cash crop, providing over 15 per cent of the world's edible oil; soy oil is an important base for margarine, cooking oils, breakfast cereals, and milk and cream substitutes, as well as being an important constituent of paints, soap etc. It is also now widely used in the preparation of meat analogues. The *groundnut* (peanut) is also an important source of edible oil. However, its requirements for a longer and warmer growing season than soy confine it to intertropical areas. The field or garden bean (*Phaseolus vulgaris*) is only one of about seventeen cultivated

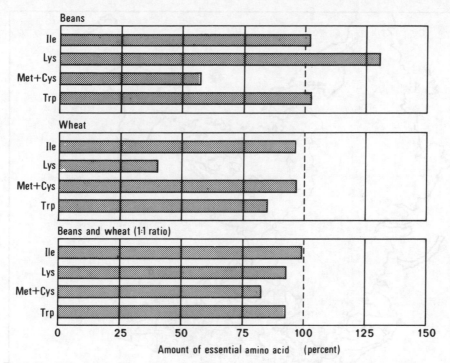

Fig. 2.2b Protein complementarity: beans and wheat. The broken line represents 100 per cent of the essential amino acid levels in a standard reference protein considered by FAO to meet human requirements (from Scrimshaw and Young, 1976)

species of bean. In intensive agricultural systems, field beans often serve the dual purpose of feedstuff and green manure. Most are warm-season annuals which require relatively high humidity and evenly distributed rainfall throughout their vegetative period.

ROOTS AND TUBERS

The final group of food crops which have become dietary staples in both temperate and tropical areas are the roots and tubers (see Fig. 2.3). Of these, the *white potato*, one of the most efficient starch-producing plants, is a basic item of diet in nearly all cool-temperate countries. Adapted to a cool growing season, it also requires a uniform moisture supply during tuber formation, and light well-drained soils. High, particularly night-time, temperatures can put it at risk from fungal infections such as the 'blight' (*Phytophthora infestans*) which decimated the Irish potato crop in successive years in the mid-nineteenth century. In addition, vegetative propagation makes the potato very susceptible to disease. As a result commercial production of seed and ware potatoes is usually spatially divorced; the former are frequently grown on the northern limits of the crop's range outside that of its main pests and pathogens. Potatoes are by far the highest yielding food crop, but, with a water content of 78 per

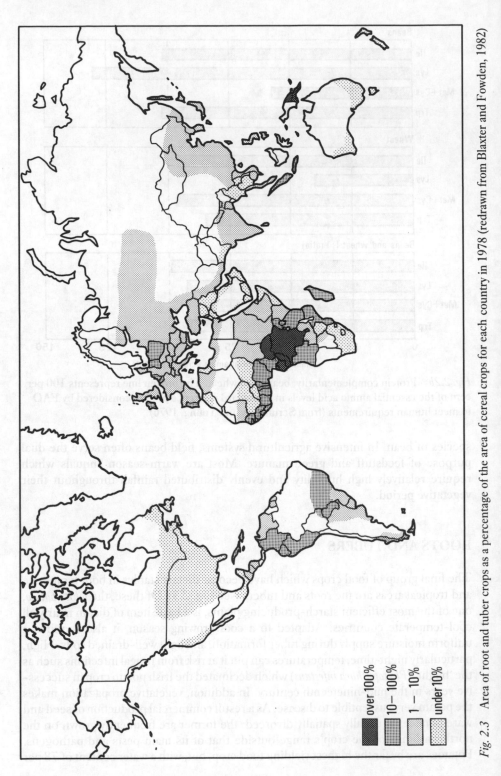

Fig. 2.3 Area of root and tuber crops as a percentage of the area of cereal crops for each country in 1978 (redrawn from Blaxter and Fowden, 1982)

over 100%

10 – 100%

1 – 10%

under 10%

cent deterioration can be rapid after several months storage. In temperate areas other root crops, particularly the swede and the turnip, have long been traditional winter feed crops.

Sweet potato and *cassava* are the most widespread of the many tropical tubers. In the humid Tropics, with a year-long growing season, and where food storage is particularly difficult, there is a high dependence for food in the traditional subsistence economies on these and other starchy tubers including yam, dasheen and taro. All need a long period of high day and night temperatures for the development of a good crop: sweet potato and cassava, at least 4 months; yam and taro, 8–12 months.

The only other food crop in the 'top twenty' is *sugar cane*. The use of sugar in large quantities in temperate countries is fairly recent; but within the short space of a century its status has changed rapidly from that of a luxury commodity to a much used commodity. It is an important source of food energy and flavouring, with a range of byproducts of which alcohol and oil are among the most important. The relatively recent domestication of sugar cane has been facilitated by rapid technological advances in the cultivation and processing of the crops. In particular, breeding for disease resistance to sugar-cane mosaic virus made its economic production possible within its present area of cultivation. A strictly tropical plant, sugar cane usually requires about 12–14 months to reach maturity, after which it is left standing, as long as temperature conditions are suitable, in order that a sugar content as high as possible can be assimilated. It requires abundant rainfall during growth, but high amounts of sunshine during the sugar-storing period. *Sugar beet* is similarly a modern crop (dependent on the development of sugar-extraction techniques), adapted to environmental conditions comparable to those of the white potato. Long days and abundant sunshine, however, are required for the production of a high sugar content. It can tolerate saline and alkali soil conditions better than most other field crops.

The only industrial crop which rivals the major food crops in extent is *cotton*, the most important fibre-producing plant (others include flax, hemp, sisal). It is a long-domesticated cultivar in South and Central America, Africa and Asia. Different types are distinguished on the basis of length of fibre (staple); of which the most important are:

Upland American (*Gossypium hirsutum*): short staple
Sea Island (*G. barbadense*): long staple
Egyptian cotton (*G. herbaceum*): long staple
Asiatic cotton (*G. arboreum*): short staple

Upland and Egyptian are now the most important species cultivated in both the Old and New World, for both fibre and for the oil-rich seed. Successful cotton cultivation requires an average temperature of at least 21–22 °C, a minimum of *c.* 500 mm rainfall during a long growing season, and freedom from frost. Originally a perennial shrub with indeterminate flowering, it is now grown as an annual. Breeding has been directed towards disease resistance and growth forms amenable to mechanized harvesting. Cotton is essentially a dual-purpose plant and its seed (with 18–24 per cent oil and 16–20 per cent protein content) is an important source of edible oil, while after extraction the remains of the seed are made into 'cattle cake'. Plant sources of oil and feedstuff, other than those already discussed, include safflower, castor bean, sunflower, hemp, linseed, sesame and rape-seed (colza). Also, the fruit of many tropical trees and subtropical palms such as the coconut and oil palm and the olive provide local or, under plantation cultivation, larger scale commercial production of oils.

While many of the food crops reviewed are dual-purpose in that they are potential sources of human food *and* of animal feed, there is a large group of plants cultivated solely for forage and which, in total, occupies very considerable areas of agricultural land. This group includes the cultivated *forage grasses* and *legumes*, i.e. clover, vetches and alfalfa (or lucerne), particularly of cool temperate areas which are used for direct grazing, ensilage or hay and as a 'break' (or rest) crop in arable crop rotations. It is, however, less easy to estimate the relative importance of the forage crops in terms of production or value because of the difficulty of recording these parameters for a crop much of which is eaten as it grows or is stored for later consumption on the farm. Even when dried and highly concentrated, the resulting material is bulky and low in value making long-distance transport except in certain circumstances generally uneconomic.

The remaining crop plants comprise *vegetables* and *fruit*; the *perennial shrubs* of which tea and coffee are the most important; and a variety of *trees*. Each of these groups contains crops grown on a small scale (as specialized garden or market-garden produce), as large-scale general-purpose field crops, or as orchard and plantation tree crops in the case of the woody perennials. At one end of the spectrum, agriculture passes into horticulture and, at the other end, into silviculture (or forestry), both of which are outside the scope of this book.

3

Agroclimate

In agricultural terms, the crop plant is the basic farm production unit, while the atmosphere and the soil provide the essential resources of carbon dioxide, water, mineral nutrients and solar radiation (light energy) on which the process of photosynthesis or primary biological production depends. All plants vary in their physical requirements for maximum performance and in their range of tolerance of physical conditions beyond which growth and/or development is limited. Climate – the average long-term atmospheric condition – is, on a world scale, the most important 'limiting factor'. While the wild or uncultivated plant requires climatic conditions that will ensure successful regeneration, the crop plant, or cultigen, needs those which will allow a high probability of a satisfactory and relatively consistent production (or yield) of the harvestable part or parts for which it has been specifically cultivated. In commercial agriculture a 'satisfactory' yield is that quantity and/or quality of harvestable product which gives maximum financial return; in subsistence economies it is that which provides sufficient food to last from one harvest to the next.

CLIMATE AND CROP GROWTH

The amount and rate of growth in any plant is dependent on the difference between gross and net photosynthesis – the process during which complex organic compounds (carbohydrates) are formed. These provide the energy for organic metabolism (growth and development) and the building blocks from which protein and fats are eventually synthesized. Provided there is no shortage of water, the closely interrelated factors of temperature and light determine the rate and amount of crop growth; and this relationship has been the basis of a long-established empirical distinction between crops of tropical, subtropical, temperate and cool-temperate regions of the world. More recently it has been shown that this zonal distribution is related to differences in the biochemical processes involved in photosynthesis – or in what are known as the particular *photosynthetic pathways* of particular crops. The three basic pathways distinguished are the Carbon-3 (C3; Calvin cycle), the

Carbon-4 (C4) and the Crassulacean Acid Metabolic (CAM). The photosynthetic characteristics of cultivars following these three pathways are summarized in Table 3.1 and Table 3.2 gives selected examples of common cultivars in each. On the basis of the operative range of tolerance of temperature and of light intensity, cultivars fall into five main groups. I and II (C3 species), III and IV (C4 species) and V (CAM species); ranges are higher in II than I and in IV than III. As indicated in Table 3.1, at high temperatures and light intensities, C4 cultivars are more photosynthetically efficient and use water much more efficiently than C3 cultivars; maximum growth rates per unit leaf area in C4 species can be twice to three times greater than those in C3 species. The latter include the temperate cereals, the white potato and rice; the former the tropical cultivars such as sorghum, maize and sugar cane. However, maximum growth responses are dependent, in both C3 and C4 cultivars, on the presence of favourable environmental conditions. C3 cultivars give maximum growth under cooler temperature and lower light intensities than do C4 species. However, under the optimum conditions for C3 species the difference between the growth rates of C4 and C3 species is relatively small.

The geographical ranges of the Group II and Group III cultivars tend to overlap. Given high light intensities Group IV (C4) cultivars, such as maize, can produce high yields in temperate continental climates with high summer sunshine amounts and temperatures. Similarly Group II (C3) cultivars, such as rice, can give high yields in tropical regions which, with a monsoon climate, combine high light intensity and temperature during the grain-filling growth-phase at the beginning of the dry season. The C3 cultivars with generally lower optimum temperatures for growth than C4 cultivars do not yield well, if at all, under sub-tropical conditions, while the latter with higher threshhold temperatures for growth (10–15°) are precluded from cool temperate areas.

The final group comprises CAM species which are adapted to grow successfully in water-deficient habitats. While they have a photosynthetic pathway and a temperature response similar to Group C4 species, they are also able to absorb and store carbon dioxide during the night-time (when their stomata remain open) and to photosynthesize during the day (when their stomata close). Species using the CAM pathway are known to occur in several families of those succulent plants so well adapted to warm and hot arid environments with large diurnal temperature ranges. There are, however, only two cultivars of agricultural importance – sisal and pineapple.

For each phase in the growth of a particular cultivar there is a temperature range within which growth and development can take place. The cardinal or threshold temperatures are: the *minimum*, below which there is insufficient heat for biological activity; the *optimum*, at which the rates of metabolic processes are at their maximum; and the *maximum*, beyond which growth ceases (still higher temperatures may be harmful or lethal). Yao (1981) gives cardinal temperatures for cool-season cereals as 0–5 °C, 25–31 °C and 31–37 °C; and for warm-season cereals 15–18 °C, 31–37 °C and 44–50 °C. Further, some crops are thermoperiodic in their response to temperature in that they require an alternation of low night-time with higher daytime temperatures for successful growth and development. Others need a degree of winter chilling (i.e. vernalization or bringing into the 'spring state') before flowering and seed-setting can take place within the growing period available.

Table 3.1 Photosynthetic pathways and related characteristics of the main crop groups

Crop group	I	II	III	IV	V
Photosynthetic pathway*	C3	C3	C4	C4	CAM
Temperature response (°C)					
optimum	15–20	25–30	30–35	20–30	25–35
operative range	5–30	10–40	15–45	10–35	10–45
Radiation intensity at max. photosynthesis (cal cm^{-2} min^{-1})	0.2–0.6	0.4–0.8	1.0–1.4	1.0–1.4	0.6–1.4
Max. net rate CO_2 exchange at light saturation (mg dm^{-2} h^{-1})	20–30	40–50	70–100	70–100	20–50
Max. crop growth rate (mg dm^2 h^{-1})	20–30	30–40	30–60	40–60	20–30
Water-use efficiency (g g^{-1})	400–800	300–700	150–300	150–350	50–200

* C3, Carbon–3; C4, Carbon–4; CAM, Crassulacean Acid Metabolic.

(from FAO, 1978)

Other crops are photoperiodic in that day-length is the 'trigger factor' necessary to initiate flowering. Four groups are normally recognized:

1. *Short-day* (or long-night), with a photoperiod of less than 10 h (e.g. soya bean, sweet potatoes, millet).
2. *Long-day* (or short-night), with a photoperiod of over 14 h (e.g. small grains, timothy, sweet clover).
3. *Intermediate-day*, with a photoperiod of 12–14 h and with inhibition of reproduction either above or below these levels.
4. *Day-neutral*, unaffected by variations in day-length.

Long-day crops will be limited to high latitudes, while the cultivation of short- and intermediate-day crops may be restricted to low latitudes and high latitudes only where spring or autumn seasons are warm enough to allow their harvest cycle to be completed. However, where temperature and/or light conditions are suboptimal for seed production, vegetative growth may be sufficient to make cultivation for fodder economically worth while.

Climatic suitability for cultivation is also dependent on the phenological or growth-phase requirements of the particular cultivar (see Fig. 3.2). Each develops through a sequence of obligatory stages or phases (i.e. germination; bud-setting; blooming; fertilization; fruiting, maturity), each of which requires a particular range and duration of environmental conditions for its initiation and completion. The length of the *phenological* or *crop growing period* is genetically determined, though

Table 3.2 Growth characteristics of harvested part of photosynthetic crop groups I–IV

Crop	Days to maturity/growing period*	Harvested part	Main use	Growth habit	Life-span cultivated	Yield location	Formation period	PS†	T‡
Group I (C3)									
Potato (*Solanum tuberosum*)	90–110 (EC) 120–140 (MEC) 150–180 (LC)	tuber	vegetable	ID	A	US	MA	V	CHT
Rape (*Brassica napus*)	100–110 (SC)	seed	oil	D	A	TI	LT	LD/DN	
Sunflower (*Helianthus annus*) (TEC)	100–120 (UBC, EC)	seed	oil	D	A	TI	LT / LP	SD/V	
	130–160 (BC, LC)			ID	A	TI	{ LT / LP }	{ LD/DN / LD }	CHF
Barley (*Hordeum vulgare*)	100–130 (SC) 180–240 (WC)	seed	grain (C)	D	A	TI	{ LT / LP }	{ LD/DN / LD }	CHTI
Bread wheat (*Triticum aestivum*) (TEC)	100–130 (SC) 180–240 (WC)	seed	grain (C)	D	A	TI	{ LT / LP }	{ LD/DN / LD }	CHF/CHTI
Oats (*Avena sativa*)	100–130 (SC) 210–270 (WC)	seed	grain (C)	D	A	TI	{ LT / LP }	{ LD/DN / LD }	CHT
Rye (*Secale cereale*)	110–130 (SC) 210–270 (SC)	seed	grain (C)	D	A	TI	{ LT / LP }	{ LD/DN / LD }	CHF/CHTI
Sugar beet (*Beta vulgaris*)	160–240	root	sugar	D	A/N	R	MA	LD	CHF

Group II (C3)									
Groundnut (*Arachis hypogaea*)	90–110 (SBC) 120–140 (ABC)	seed	grain (L), oil, cake	ID	A	LI	ME	DN	–
Soya bean (*Glycine max*) (TRC)	90–130 (EC) 140–180 (LC)	seed	grain (L), oil, cake	D ID	A	LI	ME	DN SD	–
Sunflower (*H. annus*) (TRC)	100–120 (UBC, EC) 130–160 (BRC, LC)	seed	oil	D	A	TI	LT	SD/V	–
Rice (paddy) (*Oryza sativa*) (PR)	100–130 (EC) 140–180 (LC)	seed	grain (C)	D	A	TI	LT	DN	–
Cotton (*Gossypium hirsutum*)	130–140 (OLC) 160–180 (NLC)	{ seed cotton	fibre, oil, cake	ID	A	LI	ME	SD/DN	–
Cassava (tapioca) (*Manihot esculenta*)	180–270 (EC) 270–365 (LC)	tuber	tuber, starch	ID	(SP)A	R	MA	SD	–
White yam (*Dioscorea rotundata*)	210–240	tuber	tuber	ID	(SP)A	R	MA	V	–
Oil palm (*Elaeis guineensis*)	330–365	seed	oil, cake	ID	(P)A	LI	SG	DN	–

Table 3.2 continued

Crop	Days to maturity/ growing period*	Harvested part	Main use	Growth habit	Life-span cultivated	Yield location	Formation period	PS†	T‡
Group III (C4)§									
Common millet (*Panicum milliaceum*) (TRC)	70–100	seed	grain (C)	D	A	TI	LT		–
Sorghum (*Sorghum bicolor*) (TRC)	90–110 (EC) 120–130 (MEC) 140–240 (LC)	seed	grain (C)	D	A	TI	LT LT LP	DN DN/SD SD	–
Maize (*Zea mays*) (TRC)	80–100 (EC) 110–130 (MEC)	seed	grain (C), oil	D	A	TI	LT	DN	–
Sugar cane (*Saccharum officinarum*)	270–365	stem	sugar	D	A/B	S	MA	SD/DN	–
Group IV (C4)									
Common millet (*Panicum milliaceum*) (TEC, TRHC)	80–100	seed	grain (C)	D	A	TI	LT	DN	

Crop	Days to maturity							
Common sorghum (*Sorghum bicolor*) (TEC, TRHC)	110–130 (EC)	seed	grain (C)	D	A	TI	LT	DN
	140–160 (MEC)			D	A	TI	LT	DN/SD
	170–280 (LC)			D	A	TI	LT	SD
Maize (*Zea mays*) (TEC, TRCH)	110–130 (EC)	seed	grain (C)	D	A	TI	LT	DN
	140–160 (MEC)			D	A	TI	LT	DN
	170–200 (LC)			D	A	TI	LT	DN
Group V (CAM)								
Sisal (*Agave sisalana*)	270–365	leaf		B	LE	MA		–
Pineapple (*Ananas comosus*)	365	fruit		SP	TI	LT	SD	Q

* Days to maturity for annuals; growing period required for biennials and perennials.
† PS, photoperiodic sensitivity of flowering.
‡ T, specific temperature constraint/requirement.
§ Other crops in Group III include: Japanese millet ('hungry rice' – not *Oryza*), foxtail millet, finger millet and pearl millet.

A, annual; ABC, alternatively branched cultivar; B, biennial; BC, bush cultivar; BRC, branched cultivar; C, cereal; CHF, chilling required for flowering; CHT, chilling required for tuberization; CHTI, chilling required for tillering; D, determinate; DN, day-neutral crop; EC, early cultivar; ID, indeterminate; L, legume; LC, late cultivar; LD, long-day; LE, leaf; LI, lateral inflorescence; LP, last phase in crop's life; LT, last third period in crop's life; MA, much or all of the period of crop's life; ME, middle to end period in crop's life; MEC, medium cultivar; N, normal; NLC, normal leaf cotton; OLC, okta leaf cotton; P, perennial; Q, quality of yield affected by temperature; R, root; RH, rhizome; S, stem; SBC, sequentially branched cultivar; SC, spring cultivar; SD, short-day; SG, smaller or greater fraction of plant's life; SP, short term cultivar; TEC, temperate cultivar; TI, terminal inflorescence; TRC, tropical cultivar; UBC, unbranched cultivar; US, underground stem; V, variable; WC, water cultivar.

(from FAO, 1978)

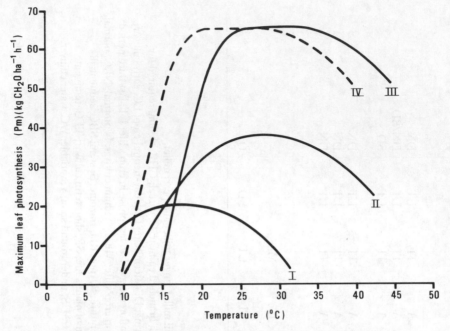

Fig. 3.1a Average relationships between maximum leaf photosynthesis rate and temperature for crop Groups I, II, III and IV (from FAO, 1978)

within it the growth phases or cycles may be shortened or lengthened by variation in weather conditions. In this respect some crops have a *determinate growth habit*, i.e. flowers (plus fruit and seeds) are borne on terminal shoots, as in the cereals and grasses; and once the apical bud becomes reproductive no more leaves are formed and the plant enters its mature phases irrespective of environmental conditions. Other crops are of an *indeterminate growth habit* in that flowers and fruit are produced, often sequentially, on lateral branches. In these circumstances the apical bud remains in a vegetative state and leaf formation is continuous throughout the plant's life-cycle and photosynthesis continues until the end of the climatic growing season.

The understanding of crop development in relation to climatic factors has made the application of *scheduling techniques* increasingly important for phasing the planting and harvesting of annual crops so as to make the most efficient use of the time and space available. Farmers once used 'guide plants' in the scheduling of sowing and harvesting. In the USA, for instance, the first English settlers learned that the time to plant corn (maize) was when the white-oak leaves were the size of squirrels' feet or when the dogwood began to show white on its wood (Wang, 1972). Scheduling is particularly important in the production of fruit and vegetable crops for the chilled or frozen food market. Thornthwaite was one of the first climatologists to construct climatic calendars for the planting and harvesting of peas for freezing by the Seabrook Farms Company in the USA in the 1950s (Wang, 1972). This involved the calculation of the number of *growth units* (i.e. accumulated average daily solar energy units) required between germination and maturity. Attempts have also been made to identify *crop response* (and hence agricultural seasons) to the

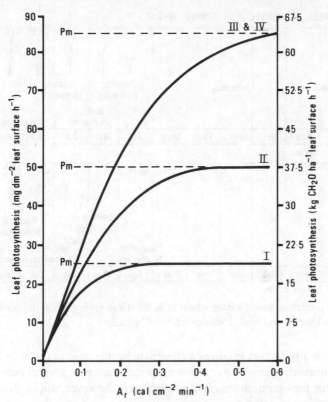

Fig. 3.1b Relationship between leaf photosynthesis rate (P_1) at optimum temperature and photosynthetically active radiation (A_r) for crop Groups I, II, III and IV. P_m is the maximum leaf photosynthesis rate at light saturation (from FAO, 1978)

distribution of daily maximum and minimum temperatures over weekly periods, for various field crops. For example Newman and Wang (1959) defined the following 'agricultural seasons' in Southern Indiana:

1. *Winter.* Crop plants dormant; more than 20 per cent daily mean temperatures 8.5 °C or less.
2. *Spring.* Early: cool-season perennial crops begin growth, and annuals (e.g. oats) planted; 20 per cent minimum temperatures less than 8.5 °C; ends when 10 per cent are 0 °C or less. Late: warm-season crops (dent corn) planted; cool-season crops grow rapidly; begins when less than 10 per cent daily minimum temperatures 0 °C or less and ends when 5 per cent or less daily minimum temperatures are 5 °C or less.
3. *Summer.* Warm-season plants (soya bean) grow rapidly; cool-season small grains (annuals) harvested; 5 per cent daily minimum temperatures 5 °C or less and 20 per cent daily maximum as low as 70 °C.
4. *Autumn.* Cool-season crops (winter grains) planted and main warm-season grains mature rapidly.

GROWTH STAGES OF WINTER AND SPRING WHEAT

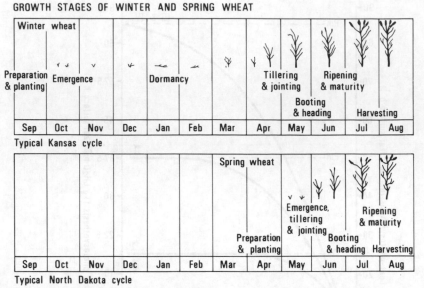

Fig. 3.2 Typical growth stages of winter wheat in Kansas and spring wheat in North Dakota (from MacDonald and Hall, 1980, courtesy of NASA)

Successful crop production requires a close synchronization of the crop growth cycle with the climatic regime. A particular crop may comprise several cultivars of varying life-spans; the optimum length of growth cycle, however, will be that which allows the production of sufficient vegetative growth to support either *continuously* (as in the case of seed-legumes, roots and other indeterminate crops) or *subsequently* (as in the determinate cereals) the necessary yield-forming activities within the climatically determined growing season.

The *climatic growing season*, or what Duckham (1963) has called the 'geographical growing season', can be very generally defined as the period (in days, weeks or months) of the year when the average combined precipitation and temperature regime permits crop growth and development (see Fig. 3.3). This concept of the growing season is an agricultural one, based on an empirically established relationship between climate and crop growth and on the recognition of heat and moisture as the two main limiting factors involved. Considerable effort and no little ingenuity has been expanded on the quantitative definition of these limits.

The *thermal growing season* has been universally defined as the length of time when mean average daily temperatures are continuously above a stated *threshold* or *base* temperature. A mean temperature of 6 °C is the longest used parameter and continues to be the commonest. It was originally chosen because of the assumption that this was the *average threshold* for the commencement of growth in temperate cereals and grasses. It is now known that individual crops can deviate widely from this threshold, while it is of little relevance in tropical areas of the world, where many crops require base temperatures for growth well above this level. In addition, it takes no account of temperatures above the optimum which could directly or indirectly curtail growth. The US Bureau of Agriculture has long made a distinction between

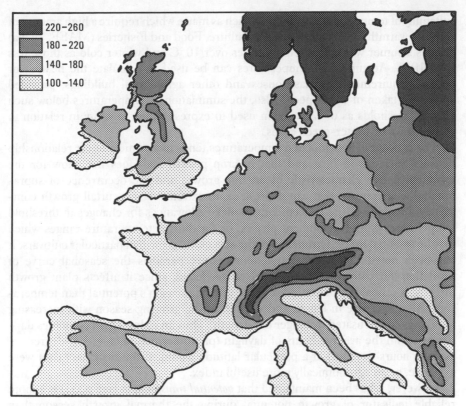

220–260
180–220
140–180
100–140

Fig. 3.3 Duration of vegetative period in Europe: number of days between seeding of summer grains in spring and of winter wheat in autumn (redrawn from Seeman *et al.*, 1979)

the *vegetative season*, as defined above, and the number of days between the average date of the first and that of the last *killing frost*. The latter is considered more important in defining the limits of economic growth of high-value, but frost-sensitive, crops such as cotton and citrus fruits. A killing frost is, for this purpose, defined as a frost of sufficient severity to damage over 50 per cent of the natural vegetation of the area. In view of the difficulty of standardizing the use of this parameter, it has now been abandoned in favour of the use of known critical minimum temperatures above or below 0 °C for particular crops and crop phases.

The relationship between the amount of heat and plant development was first quantitatively investigated by Réaumur (the inventor of the alcohol-based thermometer) in 1735 by summing the mean daily temperatures during each plant development phase. While this did not prove particularly significant, it laid the foundations for the concept of *accumulated temperatures* as a measure of the relative warmth of a growing season of given length. The remainder or summation index was calculated by summing the mean daily temperatures above the chosen threshold and expressing them in number of *day–degrees* (or heat units). Again, the threshold temperature most commonly used for general comparative purposes is 6 °C, though others have been selected for particular purposes: e.g. 10 °C is used to assess the suitability of temperate areas for crops such as maize which require a high amount of summer warmth. The Ministry of Agriculture, Food and Fisheries (MAFF, 1976a) defines summer warmth as temperatures over 10 °C and winter cold as those less than 10 °C. Accumulated temperatures can be used to calculate the heating or cooling requirements of glasshouses and other agricultural buildings, provided account is taken of wind speed. Also, the summation of temperatures below such critical thresholds as 0 °C has been used to express winter severity in relation to winter-sown or winter-green crops.

The concept of accumulated temperatures tends to assume a linear relationship between increase in heat and that in crop growth and does not allow for the exponential rate characteristic of organic growth and the occurrence of supra-optimal temperatures that can depress or, in extreme cases, curtail growth completely. In addition, the concept fails to make allowances for changes in threshold temperatures with development phases, or for diurnal temperature ranges which may be essential for or detrimental to the successful growth of particular cultivars. It has been noted that accumulated temperature parallels the seasonal curve of insolation but with a smaller annual range. Light, since it affects plant growth directly, is probably a better index of the growing season's potential than temperature. Also, decrease in length and warmth of the growing season with increasing latitude is compensated by longer day-length. Thus an index which combines day–degrees with the average hours of daylight (photothermal units = day–degrees × average hours daylight) at a particular latitude would, if the necessary data were available, be an agronomically more useful index.

However, it has been maintained that *potential transpiration* (see p. 33) is a more reliable indicator of growth potential during the thermal growth season than accumulated temperature because it is based on the direct input of solar energy. Summation of the daily 'effective potential transpiration' (i.e. when the water deficit in the main soil rooting zone is less than 50 mm) gives a good relative expression of the potential for agricultural grass production in particular.

The use of standard climatic data to describe the growing season suffers from two disadvantages. The first is that they do not describe the condition of the atmosphere surrounding the plant (i.e. the microclimate); most meteorological parameters are recorded at a height well above the general level of the majority of field crops. The second is that there can be an enormous difference between the climate of the root and that of the shoot. Consequently *soil temperature* is now considered of greater agronomic significance than is air temperature in the influence it exerts on such activities as germination and root development, and hence on growth above and below the soil surface. Soil temperature can (if data are available) give a more acceptable and realistic measure of the length and intensity of the growing period than can free-air temperature. It has been suggested that an acceptable measure of the *grass growing season* in England and Wales (MAFF, 1976a) can be deduced from the length of time the earth temperature remains above 6 °C at 30 cm under a closely cropped grass sward. At this depth diurnal variations of temperature at the recording time of 09.00 h are negligible. The soil horizon above this depth is omitted in order to avoid the effects of varying soil type and land use on the temperature regime. It must, however, be noted that the 'growing' and the 'grazing' season are not necessarily coincident, since grass must be allowed to grow to a height that will allow sufficient feeding without detriment to the sward. Further, the end of the grazing season is not easily defined since it may depend on a number of factors other than grass growth. In addition, *accumulated soil temperatures* (or the 'T-sum') can be used as an indicator of when to apply inorganic fertilizers (particularly nitrogen) in order to achieve as close a synchronization as possible with crop growth in spring.

THE HUMID GROWING SEASON

The length and effectiveness of the growing season is also dependent on the availability of sufficient soil water to initiate growth and ensure the successful development of the harvestable product. In many areas of the world, water is the main factor limiting crop cultivation.

Water availability for the plant is dependent on the amount of effective rainfall which enters and is retained by the 'soil reservoir'. A satisfactory quantitative expression of the humid growing season has proved more difficult than of the thermal season. There is a paucity of direct measurements of either evaporation, transpiration or evapotranspiration. Also, the relationship between evaporation and actual or potential evapotranspiration, as well as that between soil moisture levels and plant growth is difficult to establish precisely; this is an area of continuing debate. Livingstone (1916) was one of the first workers to devise a *hydrothermal index* (for the purpose of defining and mapping climatic zones in the USA) which used a 'temperature efficiency index' derived from observations of the rate of growth of maize shoots under varying temperature, precipitation and evaporation. However, its application was limited by a lack of evaporation measurements. Records of temperature or saturation deficit are more readily available than those of evaporation and were initially used to express the relationships between precipitation and evaporation and to formulate indices of relative aridity or humidity. Workers at the Waite Institute in Australia proposed a precipitation to evaporation (expressed by saturation deficit) index (PE) of 0.5 called the Waite index, for the beginning of plant growth or the so-called 'break of season'. However, it was realized that continuous growth would require a higher ratio if the rate of transpiration was high and a lower ratio if the rate was low. Subsequently $PE^{0.7}$ was recommended as a measure of available moisture and an index of 0.4 as the threshold for growth. The growing season was then deemed, for all practical purposes, to be those months when $PE^{0.7}$ exceeded 0.4 and the mean temperature was over 2 °C (see Fig. 3.4).

Thornthwaite (1948) and Penman (1956) developed and applied the concepts of *potential evapotranspiration* and *potential transpiration*, respectively, to express soil water deficits or surpluses for plant growth. Potential evapotranspiration was defined as the water loss which will occur if at no time there is deficiency of water in the soil for use of vegetation (Thornthwaite, 1948). Potential transpiration was defined more fully as evaporation from an extended surface of a short green crop, actively growing, completely shading the ground, of uniform height and not short of water (Penman, 1948, 1949). In the face of the scarcity of direct measurements of

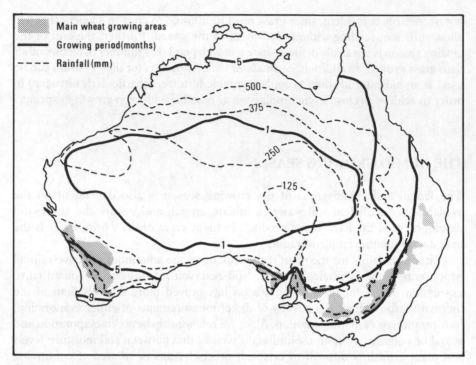

Fig. 3.4 Comparison of wheat-growing areas: mean annual rainfall (mm) and length of growing season (months) in Australia (from Jeans, 1977)

either evaporation or evapotranspiration both Thornthwaite and Penman constructed empirical formulae by which they could be estimated. The former expressed potential evapotranspiration simply as a function of mean air temperature and day-length. The latter devised a more complex but more comprehensive set of formulae which combined nearly all the factors involved (i.e. saturation deficit; temperature; wind speed; net radiation).

The concepts of potential evapotranspiration and transpiration provided a means of arriving at estimates of *actual evapotranspiration* and *transpiration* and hence of calculating soil moisture deficits, as well as allowing a more realistic measure of the humid growing season. The initial, and still frequently used, expression of climate humidity or aridity is the amount by which precipitation exceeds, equals or falls below potential evapotranspiration. This, however, fails to allow for that water stored in the soil and held by capillary forces, after all the surplus water has drained away, and which is capable of being extracted by plant roots. The maximum amount of this available soil water depends on a number of factors, not least soil depth and texture; suggested variations range from 4 cm storage capacity for each 100 cm depth of sand to 17 cm storage capacity or more in the same depth of clay loam (Ward, 1975). Thornthwaite (1948) used an average of 10 cm water storage in his water-balance calculations. Penman *et al* (1949), however, argued that the readily available water would be that in the *rooting zone* which is determined partly by soil depth and partly by soil type; this they termed the *root constant*. Its value for grass was estimated as 7–12 cm but for more deeply rooting shrubs and trees it may be 25–30 cm. In those areas of the world where moisture is the factor determining the length and effectiveness of the growing season, recent work has tended to concentrate on improving such estimates of soil moisture.

The main agricultural significance of the work by Thornthwaite and Penman has been the application of the concepts of potential evapotranspiration and transpiration, based on climatological data, to the expression of soil water balance in terms of *deficits* and *surpluses*. The soil water deficit is the amount by which soil moisture falls below *field capacity*, the amount required for maximum growth and development. A deficit of over 50 mm in the main rooting zone will impede grass growth. The surplus water present in excess of field capacity gives rise to soil saturation and waterlogging with deleterious effects on crop development. In many areas soil water deficits during the growing season can depress growth to an extent that makes supplementary irrigation economically viable. Calculation of irrigation need, however, can be problematic. It can be done on the basis of observed plant or soil responses to water deficiencies. There are instruments which can be permanently installed which respond to changes in soil moisture at a particular point. However, despite their imperfections the climatological approaches taken by Thornthwaite and, more particularly, Penman to the calculation of supplemental irrigation need have been universally adopted and are still used to estimate or predict soil water deficits on a regional scale.

In Britain, soil moisture and, more particularly, drainage conditions have been used to define the growing season, particularly where this may be curtailed by a surplus rather than a deficiency of water. Winter is defined as the period when the soil water is over field capacity (MAFF, 1976b), the average date of which will be that when field drains begin to run and after which the liability to flooding after heavy rainfall increases. Further, the amount of winter rainfall gives an indication of drainage needs, as well as of losses of lime and other nutrients by leaching. In contrast, the summer drainage period is that when soil moisture is, on average, at or below field capacity and when growth is not checked directly or indirectly by surplus soil moisture. The commencement, the earliness or lateness of the growth-favourable period will, given suitable temperatures, be dependent on either weather conditions and/or the soil moisture-holding capacity.

Palmer (1968) defined *agricultural drought* as a significant reduction of available moisture below that necessary for the 'near-normal' operation of the type of farming characteristic of a particular area. He devised a 'drought index' based on successive weekly values of computed evapotranspiration deficits. Maps of this index are now a regular feature of the *Weekly Weather and Crop Bulletin* issued by the National Oceanic and Atmospheric Agency in the USA.

There are, however, situations where the growing season may be bimodal, in the sense that there may be a humid phase, with high precipitation inputs, and a subhumid phase, with cultivation possible, though primarily dependent on stored soil water. In the recent analysis of the growing season in relation to crop suitability in Africa, a working definition of this season is given as 'the period in days during a year when precipitation exceeds half the potential evapo-transpiraion, plus a period required to evapo-transpire an assumed 100 mm of water from excess precipitation (or less if not available) stored in the soil profile, and when temperature does not fall below the minimum at which crop growth is possible' (FAO, 1978, 33).

Most measurements of the growing season on any but a very local scale must inevitably suffer from the inbuilt limitations of all agroclimatic indices. First, the stations providing recorded or observed data are often sparsely distributed, and may be absent at high altitudes. Secondly, the number of records kept often vary: those for evaporation, sunshine hours, radiation and soil temperature are much fewer than those for precipitation and temperature, as are the number of agrometeorological stations recording 'crop weather'. In addition, the availability of short-term (e.g. daily, weekly) in comparison to monthly and annual records is low. Within even a small area the actual length of the growing season may vary by a factor of weeks.

Decrease of soil temperature with altitude has been shown to parallel, but at a faster rate, that of the air temperature (average rate, 0.6 °C per 100 m) (Seeman *et al*, 1979). Variation of temperature with altitude can, however, depend on season and local topography as shown in Table 3.3: in general, the decrease in summer is

Table 3.3 Decrease of temperature extremes with every 100 m increase in altitude (°C)

	Winter comparison of stations in:			Summer comparison of stations in:		
	narrow valleys	wide valleys	open high ground	narrow valleys	wide valleys	open high ground
Maximum	−0.7	−0.7	−1.0	−1.0	−1.2	−1.4
Minimum	−0.8	−0.5	−0.6	−1.2	−0.5	−0.4

(From Seeman *et al.*, 1979)

greater than that in winter. However, on sloping terrain, cold-air drainage may be caused by or create temperature inversions, and increase the incidence of early and late frosts on lower slopes or on valley floors. Slope gradient and aspect, as a result of their effect on energy received from solar radiation, can have an appreciable influence on the length of the growing season. Measurements have shown that, in Britain, there may be as much as 20 days difference in length of growing season between a south-facing slope of around 20° inclination and a level site in the same locality. In northern latitudes the difference will be greatest and agriculturally most significant in spring and autumn, when the inclination of the sun's rays are lower, than in midsummer (see Table 3.4). South-facing slopes have the advantage of drying out and hence warming up earlier than those facing north (see Table 3.5). Also, although east-facing slopes receive almost the same incident radiation as those to the west, the latter have slightly longer periods of sunshine in the autumn. In addition, slopes can afford shelter from exposure to wind such as to make the area to the lee appreciably warmer than that to windward.

Table 3.4 Total monthly insolation from direct sunlight in cloud-free weather for slopes of various inclinations and orientations for lat. 50°N

		Total monthly insolation (kcal cm^{-2})					
		Orientation					
Month	Inclination	N —	NE — NW	E — W	SE — SW	S —	Difference between north and south slope
December	0° (flat)	2.0	2.0	2.0	2.0	2.0	0.0
	10°	0.7	1.2	2.0	3.0	3.5	2.8
	20° (slope)	0.0	0.6	2.0	4.0	5.0	5.0
	30°	–	0.1	2.0	4.8	6.0	6.0
	90° (cliff)	–	–	1.4	5.7	8.3	8.3
March	0°	8.8	8.8	8.8	8.8	8.8	0.0
	10°	7.1	7.9	9.2	10.6	11.2	4.1
	20°	4.9	5.6	8.9	11.8	12.8	7.9
	30°	2.1	4.6	8.7	12.6	14.1	12.0
	90°	–	1.2	5.6	8.8	12.2	12.2
June	0°	18.6	18.6	18.6	18.6	18.6	0.0
	10°	17.8	18.1	18.7	19.4	19.5	1.7
	20°	16.5	17.0	18.4	19.7	19.6	3.1
	30°	14.2	15.0	17.5	18.8	18.8	4.6
	90°	2.0	5.4	8.8	8.0	7.0	5.0
September	0°	10.8	10.8	10.8	10.8	10.8	0.0
	10°	9.2	9.6	11.0	12.1	12.9	3.7
	20°	6.5	8.0	10.9	12.2	14.4	7.9
	30°	3.9	6.3	10.6	13.8	15.4	11.5
	90°	–	1.8	6.3	9.9	11.2	11.2

(From Seeman *et al.*, 1979)

Table 3.5 Temperature differences at 10 cm in July between soils on northern and southern slopes at a gradient of 20–22° (°C)

	Time of day (h)			
Soil surface	10.00	12.00	14.00	16.00
Bare stripped	8.4	11.8	16.1	15.7
Grass covered	3.2	4.3	6.2	7.4

(From Seeman *et al.*, 1979)

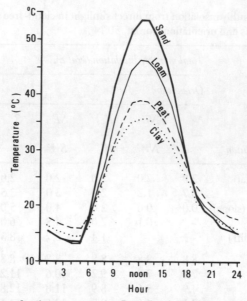

Fig. 3.5 Daily course of soil temperature at 5 cm depth on clear summer days at Sapporo, Japan (from Chang, 1968a)

Finally, soil texture and colour can cause small-scale, but quite significant, variations in the date of commencement, as well as the length, of the growing season. More important, however, than the greater albedo of light-coloured soils compared with dark-coloured soils is the soil water content (see Fig. 3.5). The thermal conductivity of air is greater than that of water but its heat capacity is much less. As a result poorly drained water-retentive soils warm up slowly in the spring in comparison to those that are less water-retentive and more freely drained. The former tend to be called 'cold soils' in that they warm up slowly in the spring but, in contrast to the 'warm soils', they retain heat longer in the autumn. In addition, by modifying the length of the growing season, land form can exercise a significant influence on the altitudinal limits of cultivation or choice of crop grown. Favourable slope and soil climates are particularly important in respect of frost-sensitive fruits, and 'early' vegetable crops for instance. They may also allow the cultivation of crops in areas or at times that might otherwise be climatically marginal but where quality or earliness more than compensate economically for the relatively lower yields that may result.

Climate affects not only crop growth but also the conditions under which cultivation takes place. Duckham (1963) made a useful distinction between the *bioclimate* and the *ergoclimate* (i.e. work climate). The latter term refers to those conditions which influence the ease or difficulty and hence the speed of farm work, and which may, as a result, curtail the length of the climatic growing season by delaying one or more of the cultivation processes. In addition, rain and wind directly affect the physical condition of crops and animals so as to make their 'handling' difficult in terms of both the time and energy expended on farm work. The effect of climate is, however, most important with respect to the *timeliness* of field operations, since there are limits to the seasons or periods within which certain operations can be carried out successfully. One of the most important limits is that of soil or land accessibility. The *accessibility period* has been defined as the number of days in the year when machinery can work on the land and animals can use pasture without detriment to the condition of the soil or sward. Accessibility is a function of the *cultivability* and *trafficability* of the soil, both of which depend on soil moisture conditions as determined by soil texture on the one hand and by climatic conditions on the other, and which will be discussed more fully in the next chapter.

While climate exercises a direct effect on the range and productivity of crops, it also operates indirectly through its influence on the geographical range, the seasonal development and the activity of crop *pests* and *pathogens* on the one hand, and on the susceptibility of the host plants to attack on the other. The three climatic factors considered most important in affecting the development and spread of pests and disease are temperature, relative humidity and wind speed. Temperature often controls regional distributions, particularly in terms of latitude, altitude and seasonal incidence. Optimum temperatures for infestation or infection and subsequent development vary with the host species and the varieties of pest or pathogen; the length of the life-cycle of such organisms decreases and the rate of their development increases with increasing temperature. Valli (1968), for instance, notes that at a temperature of $0\,°C$ wheat stem-rust develops in 3 months, at $4\,°C$ in 22 days, and at $24\,°C$ in only 5 days.

Pests and pathogens also vary in their range of tolerance of extremes of temperature. Host plants and animals may be less susceptible to attack near the margin of the pest range. For instance crops such as potatoes or grass produced for seed are often grown near the margin of their climatic range and hence beyond that of the economically destructive diseases such as blight and scab. In these cases the value of the 'crop' is related to its disease-free quality rather than to its absolute yield. Precipitation and atmospheric humidity may, alone or in combination with temperature, be a controlling factor in the distribution and incidence of pests and disease. Soil climate is sometimes regarded as more important in this respect than the condition of the free atmosphere. However, the height, density and areal extent of the crop through its effect on the microclimate and the rate of spread of pests and pathogens may be more important in tipping the balance in the struggle between the host and its attackers.

Finally, wind can cause direct physical damage to plants by sand and salt blasting, leaf tattering or plant breakage. In addition, desiccating winds can give rise to conditions of severe water stress for animals, while at the same time exacerbating the effects of abnormally high or low temperatures. Wind is also an important agent in the dispersal of pests and pathogens. Other climatic factors, such as light intensity and day-length, can affect the length of incubation periods and rapidity of development in certain pests and pathogens.

The close relationship between pest/pathogen activity and climate makes the timing as well as method of control particularly important. There has been an ever-increasing development of pest/disease warning systems on the basis of climatic prediction and meteorological forecasting – the aim of which is to provide information on the probable occurrence of outbreaks far enough in advance to allow control measures to be applied most effectively. Successful and efficient protection

Table 3.6 Lengths of wetting period (in hours) necessary at different mean temperatures to produce light, moderate or heavy infections of apple scab

Mean temperature during wetting period (°F)	Level of infection		
	light	moderate	heavy
7°C	20	26	40
13°C	12	16	24
18°C	9	12	18
24°C	12	17	26

(From Valli, 1968)

depends on speedy and accurate forecasting of the environmental conditions favourable to the onset of the pest or disease outbreak. Some of the most sophisticated methods of forecasting are associated with such particularly virulent and economically disastrous diseases as potato blight and scab (on various temperate fruits). In the latter case a disease-incidence estimate is based on the 'wetting period', i.e. the number of hours during which leaves will remain wet, and the mean temperature forecast during this period (see Table 3.6). More progress, however, has probably been made in the prediction of animal disease, where the time-lag between meteorological events and pathological manifestations is largely independent of the weather *after* the critical event (Duckham *et al*, 1976). More recently, forecasting on the basis of computer models designed to simulate the reaction of a particular pathogen to current meteorological conditions has been developed for many pests and diseases in many of the technologically advanced countries.

Because of its dependence on climate most types of agriculture are inevitably subject to the unpredictable long- and short-term fluctuations characteristic of atmospheric conditions. Climate introduces an element of risk in farming which makes it a particularly hazardous occupation. A *climatic hazard* in this context can be defined as any type of extreme atmospheric (i.e. actual weather) condition that can damage or destroy crops, animals or agricultural buildings and cause a particularly large abnormal decrease in production and/or income. The main types of climatic hazards are summarized in Table 3.7. The extent to which climatic variations create a serious hazard or risk depends on a number of variables. These include, on the one hand, the time of occurrence, the intensity or magnitude and the duration of the particular event and, on the other hand, the age, stage of development and inherent resistance (or tolerance) of the crop plant or type of livestock involved. Plants and animals are normally more sensitive to extreme conditions during the growing season and, more particularly, in their early stages of growth, while crop plants have varying critical periods or development stages when extreme conditions may seriously reduce or check production of the harvestable product.

The severity of a climatic hazard will also be dependent on its frequency and the degree of predictability of its occurrence and hence the extent to which it is physically possible and economically feasible to minimize the impact. Three strategies identified by Hudson (1977) which are designed to minimize climatic hazards include:

Table 3.7 Climatic hazards

Climatic elements	Type of hazard	Methods of protection and/or control
Temperature	frost	flooding/sprinkling; preventing heat loss from soil – mulching etc.; air circulation by wind machine
Precipitation excess rainfall (amount or duration)	flooding	
inadequate rainfall	drought	irrigation
	hail	
	snowdrifts	snow fences
Precipitation + temperature	glazed ice	
Wind	gales	shelter
	tornadoes	–
	hurricanes	–
Wind + precipitation	rain storms	
	snow blizzards	shelter
Wind + temperature	desiccating winds	shelter
	fire	fire rides/corridors; fire-resistant shelter

1. Climatic and weather modifications which are still only really successful on a small scale such as the measures to modify the microclimate in order to maintain temperatures near ground surface above freezing level; to reduce the direct and associated effects of high wind force by shelter or to mitigate drought by irrigation.
2. To change the plant response to weather by the use of exogenous growth regulators which can stimulate plants to grow better under suboptimal conditions or to grow to heights which will be less susceptible to damage.
3. To adjust management as carefully as possible to current weather conditions.

The last strategy depends on a deeper understanding of crop–weather relationships than is available at present, and on sufficient records of agronomically significant data to allow accurate timely forecasting of an imminent hazard and the probable time of recurrence. As Waggoner (1968) notes, meteorological data are needed by the farmer so that he can decide those tactics best suited to cope with current weather as well as to enable him to plan long-term strategies in relation to climatic variability. Insurance against severe unpredictable climate hazards can be exorbitantly expensive. Nevertheless, in some types of farming there are hazards of such frequency and severity (such as hail in US wheat-growing areas) that 'insurance' is an accepted cost.

The cultivated soil

Cultivation involves, on the one hand, the selection and production of those crops or crop varieties which will give as satisfactory a yield (or economic return) as possible under prevailing physical and cultural conditions; and, on the other, the management of the physical environment to produce as favourable a habitat as possible for successful crop production. That part of the physical environment most amenable to management is the soil, the cultivation of which comprises a range of processes including:

1. *Tillage*: physical preparation of the ground and soil.
2. *Sowing or planting*: initial crop establishment.
3. *Drainage and/or irrigation*: management of air and water in the soil.
4. *Fertilization*: input of nutrients.
5. *Crop protection*: weed, pest and disease control.
6. *Harvesting*: final collection or removal of end-product.

While these processes are basic to all types of cultivation, the relative importance of the methods used can vary according to environmental conditions, the nature of the crop and the techniques available.

Tillage is not only the initial process in cultivation, it has until relatively recently been the principal process in the creation of the cultivated soil. Its aims are to: create soil conditions that will facilitate seed germination, seedling emergence and root development; inhibit or destroy competitive organisms such as weeds, pests and pathogens; and allow crops to be cropped or harvested easily and in good condition. To these ends the soil is disturbed, opened up and turned over. The intensity and depth to which this is effected are usually dependent on site and soil type on the one hand, and on the type of implements used on the other; but may also be a function of the type of crop in some instances.

The digging stick and simple hoe were the earliest tools of cultivation used by man. Still widely employed throughout the less-developed parts of the world, they do little more than scratch the surface of the ground. The *plough*, however, digs into and inverts the top-soil. Developed early in the history of man, it has remained the

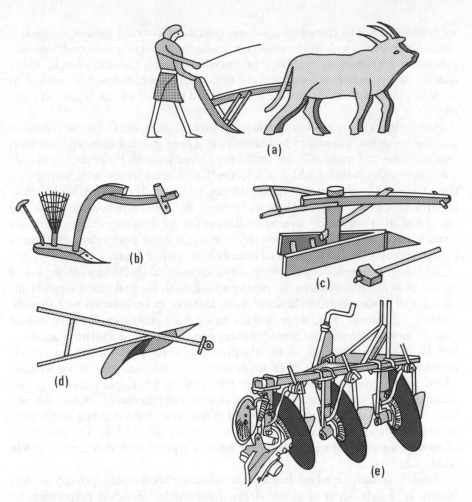

Fig. 4.1 Development of the plough: (a) Assyrian plough drawing in a tomb; (b) Egyptian plough with seed-box, *c.* 3000 BC; (c) Medieval plough with wooden mould-board, and mallet to break up clods; (d) one of the first iron ploughs with curved mould-board – the 'Rotherham plough', seen from above; (e) a modern disc plough (adapted from Sutherland, 1968)

most effective and widely used implement for breaking up the soil and burying weeds (see Fig. 4.1). The ancient Egyptians and Sumerians used heavy wooden ploughs drawn by oxen, such as are still used in many parts of India. They break up but do not turn over the surface soil. It was not until much later – in the Middle Ages – that the mould-board plough, as it is known today, came into use. This combined a knife (or coulter) to cut the sod, a share to penetrate the soil and a wooden mould-board to push the soil to one side; initially a heavy wooden hand-mallet was required to break up the clods. The latter was later replaced by the harrow (or rake).

It was not, however, until the scientific and technical developments of the eighteenth century that the range and diversity of agricultural implements began to increase rapidly. The curved ploughshare, capable of inverting the soil, was developed. Since then a wide range of ploughs, adapted to varying soil conditions and cultivation requirements, has been produced, including the disc plough, which shatters and pulverizes the soil, and the tined plough, which makes it possible to work downwards from the surface in stages and so avoid the excavation of large clods.

Among other innovations in Britain in the eighteenth century were the seed-drill and the horse-hoe introduced by Jethro Tull. These ensured more efficient seed establishment and more effective methods of weed control. However, it was only when herbicides, which could deal selectively with grass weeds, were introduced that tillage could, under certain circumstances, be reduced or omitted. With the use of herbicides, and trash control by straw burning, the main purpose of tillage became the provision of satisfactory sowing conditions for the next crop. Today there is a trend to put most effort into top-soil cultivation and to use a *cultivator* rather than a plough on heavy land or to drill seed directly into untilled land.

The other major modern development in agricultural machinery was the use of motive power. Until well into the nineteenth century, the main motive power for agricultural work was either human labour alone or in conjunction with draught animals (e.g. horse, mule, water buffalo, ox etc.) for pulling implements; indeed draught animals are still widely used in many parts of the world. Full mechanization had to await the invention of the steam-engine (first used for driving cereal-threshing equipment) and, more importantly, the internal combustion engine, which found agricultural expression in the farm *tractor*. Rapid increase in use, accompanied by even greater diversity of design and function of tractors did not, however, take place until after the Second World War. Indeed as late as 1940 the number of horses still equalled that of tractors on farms in the UK. The main types of modern agricultural machines on the basis of type of work done are set out in Table 4.1.

Traditional tillage methods in humid temperature areas involve primary cultivation to a plough depth of about 20 cm, followed by seed-bed preparation by harrowing or rotary cultivation to about 10 cm. The depth to which the soil is disturbed by tillage, however, varies. The shallow or skim plough is designed to work only to a depth of 10–15 cm. Also, there are conditions which make tillage physically difficult or even impossible. In some cases the soil material may be too shallow or too stony or the surface soil strength too weak to make tillage worth while. In other cases slopes can be too steep for any form of tillage without terracing; 15° is regarded as the average slope above which the use of the tractor-pulled plough or cultivator is hazardous – though it can range from 13° for a stand-wheeled tractor to 17° for a tractor with four wheel drive. The severity of these limiting factors has in fact increased with technical developments and with the evolution of highly specialized, often heavy and always costly, tillage implements. Indeed increasing size and cost of modern tillage implements has seen an increase in the size of area that is considered profitable to cultivate. Areas too small may be abandoned or field size increased where possible.

Whatever the method used, the aim of tillage is to produce as good a *soil condition* or *tilth* as possible for crop establishment and initial shoot and root development.

Table 4.1   Farm machinery grouped according to type of work done

1. *Prime movers*: tractors
2. *Digging, turning and burying*: mould-board ploughs, disc ploughs
3. *Deep penetration*: subsoil ploughs, scarifiers, chisel ploughs, deep tillers
4. *Loosening and weeding*: cultivators, spike harrows, disc harrows, tillers, hoes
5. *Chopping and burying*: trash cultivators, rotary hoes
6. *Digging out*: cane grubbers, deep scarifiers
7. *Compactors*: rollers, culti-packers
8. *Sowing*: drills, seeders, sod seeders, minimal-tillage seeders, combines, transplanters, seed broadcasters, aeroplanes
9. *Harvesting*: balers, pick-up balers, reapers and binders, headers, auto-headers, corn pickers, cane harvesters, cotton harvesters, potato spinners, carrot harvesters, forage harvesters, ensilage machines
10. *Crop-handling*: loaders, augers, elevators, chaff cutters, hammer mills, driers, hoppers, corn shellers, seed graders, fruit graders
11. *Fertilizer-spreading*: direct-drop spreaders, rotary spreaders, aeroplanes, liquid-ammonia drills
12. *Mowers*: lucerne mowers, flail mowers, rotary-blade mowers
13. *Earth-moving*: bulldozers, front-end loaders, scoops, graders, post-hole diggers, trench diggers
14. *Spraying*: knapsack sprays, boom sprays, orchard sprays, aeroplanes
15. *Dairying*: milking machines, separators
16. *Irrigating*: pumps, fixed sprays, movable sprays

(after Sutherland, 1968)

Table 4.2   Heavy soils: textural classes and approximate mechanical composition (%)

	Mechanical size fraction			
	coarse sand (2.0–0.2 mm)	*fine sand* (0.2–0.02 mm)	*silt* (0.02–0.002 mm)	*clay* (<0.002 mm)
Sandy clay loam	25±5	25	10±5	30±10
Clay loam	10±5	30	20±5	30±10
Silty clay loam	<5	20	35±10	40±10
Sandy clay	25±5	15	10±5	>40
Clay	10±5	20	20±5	>50
Silty clay	<5	10	35±10	>40

(from Wilkinson, 1975)

The ideal is a soil in which the porosity of the mineral matter provides an optimum balance between water-holding and freely drained, well-aerated conditions. The extent to which this can be achieved is dependent on a number of variables of which the most important is *soil texture*. Texture is an inherent soil property which it is difficult to modify on other than a very small scale. It is dependent on the relative proportion of the constituent soil mineral particles within accepted size ranges or classes. The three main soil fractions or separates by which the soil texture is described are sand, silt and clay (see Table 4.2). The agricultural significance of texture is related, on the one hand, to its effect on the *porosity* and *permeability* of the soil; and, on the other to the *surface area* of the soil fraction. The size of the interparticle spaces or pores determines the rate at which water drains through the soil. The surface area of the constituent particles determines the amount of water and nutrients in solution that can be retained against the force of gravity. As Fig. 4.2

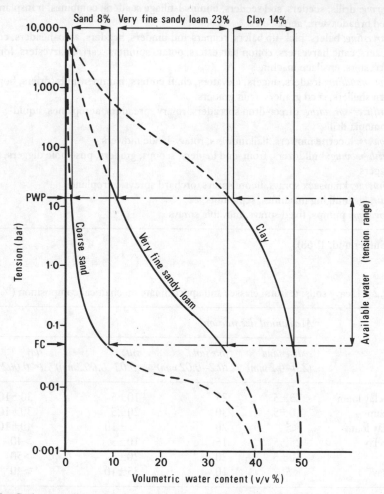

Fig. 4.2 Soil water characteristic curves: FC = field capacity; PWP = plant wilting point (from MAFF, 1982b)

illustrates, a heavy soil can hold more than twice the volume of water than can a light one. The traditional categorization of agricultural soils as *light*, *medium* and *heavy* is a description primarily of their *workability* as affected by texture. The light soils (with over 80 per cent sand) are coarse-textured. They are usually described as dry, 'droughty' and 'hungry' because of the rapidity with which water percolates through them and nutrients are leached from them. They have, however, the advantage of warming up more rapidly and allowing earlier growth in spring than the heavier soils. In contrast, the heavy soils are fine-textured, comprised usually of over 25 per cent clay. Because of the high proportion of micropores (i.e. pores of diameter less than 0.002 mm and, more particularly, less than 0.001 mm) and the very large, chemically active, surface area of the constituent clay particles, these soils are both water- and nutrient-retentive. In addition, some of the clay particles can imbibe water, and hence swell when wetted and shrink when dried. Because it is chemically active, clay tends to dominate the other soil fractions and hence to exert a physical and chemical influence out of proportion to its amount in the soil. The clay fraction, however, is usually composed of a mixture of clay mineral (one of which may be dominant) whose physical and chemical properties vary widely (see Table 4.3).

Table 4.3 Properties of clay minerals

Clay mineral	Al : Si ratio	Other elements	H-bonds	Cation exchange capacity
Montmorillonite	2 : 1	Fe, Mn, Ca	weak	high
Vermiculite	2 : 1	Fe, Mn, Ca	weak	high
Illite	2 : 1	K, Ca	weak	high
Kaolinite	1 : 1	–	strong	low
Allophane	1 : 2 (amorphous)	–	–	high
	Fe : Al ratio			
Hydroxide clays	1 : 2	–	–	very low

H = hydrogen; Al = aluminium; Si = silicon; Fe = iron; Mn = manganese; Ca = calcium; K = potassium.

SOIL STRUCTURE

One of the primary aims of soil management is to create as good a soil condition or *structure* as possible for crop growth and thereby enhance the advantages and mitigate the limitations of the available soil texture. The structure of the soil is a function of the way in which the individual mineral particles are aggregated into compound particles or *peds*. The latter can vary in size, shape, arrangement and stability depending on a number of interacting variables of which the most important are the initial soil texture, the organic material content, physical processes and soil management. That structure considered ideal for agriculture is described as '*crumb*', in which the peds are small, porous and relatively water-stable. However, the shape and structure of the peds *per se* are now considered less important from the agronomic standpoint than are the size and continuity of the air spaces between and within them (see Table 4.4). Indeed *porosity* is now regarded as the best measure of the soil's structural condition. Hall *et al* (1979), for example, have ranked structural quality in the top-soil on the basis of the air capacity and available water as follows:

	Air capacity (%)	Available water (%)
Very good	>15	>20
Good	10–15	15–20
Moderate	5–10	10–15
Poor	<5	<10

However, for optimum structure a range of pores of varying size is required to allow ease of root penetration, free drainage and adequate water storage. Hence the ideal structure is one with abundant pores and fissures of over 0.1 mm diameter which permit free root growth, oxygen diffusion and water movement, combined with those of less than 0.05 mm which are capable of retaining water against the force of gravity. In addition, infiltration of precipitation should be rapid enough to prevent the accumulation of surface water; compaction in the subsoil should be absent and rooting should be to a depth of a metre in some cases, if possible.

The development of soil structure results from the complex interaction between weather, cultivation methods and, in the first instance, soil texture. Sandy and silty soils do not aggregate easily, while those with a high proportion of clay are highly cohesive, swelling on wetting, and shrinking and cracking along vertical fissures to give massive blocky structures when dry. A certain proportion of clay, however, forms one of the essential binding agents necessary for the aggregation of sand and silt particles into crumbs. Of even greater importance in the production of stable soil structures is the presence of well-decomposed organic matter, which, like clay, has colloidal properties and which, under certain circumstances, occurs in close association with clay to form the soil *colloidal complex*. The water- and nutrient-holding capacity of humus is considerably greater than that of clay. Its presence is essential for structural stability. In many agricultural soils, particularly those continuously cropped, humus levels may be nearly at the critical limit for structural stability (4 per cent in fine sandy and silty soils, 2.6 per cent in heavy soils) if the only source of fresh organic matter is that supplied by crop roots and other residues (see Fig. 4.3). Synthetic substitutes, such as Krillium, have been produced but are less effective and are usually too costly, except for use on a very small scale.

Other factors contributing to the development of a good structure are the soil micro-organisms, which produce sticky substances and eventually contribute to the store of dead organic matter in the soil. In temperate areas of the world, Enchytraed worms play a vital role in the breakdown and intermixing of organic with mineral matter. In addition, they form channels 20–50 mm in diameter, which can act as storage pores. Finally, grass roots are very important. Not only do they contribute a continuous supply of organic matter directly into the soil but, it has been suggested, the densely ramified network of fine roots (particularly of many agricultural grasses) appears actively to promote a crumb structure (see Fig. 4.4). This may be a result partly of the fine subdivisions they create within the soil and partly of the activity of micro-organisms in the root zone itself.

Table 4.4 Functional descriptions for pore size groups*

Greenland (1977)		De Leenheer (1977)		
Equivalent pore diameter	Functional description	Equivalent pore diameter	Functional description	pF†
>500	fissures	>300	aeration capacity	0–1
500		300		
	transmission pores		normally draining pores	1–2
50				
		30		
			slowly draining pores	2–2.54
	storage pores			
		9		
			useful water-retention capacity	2.54–4.19
0.5		0.2		
	residual pores		non-useful water content	
0.0		0.0		

* The divisions are necessarily somewhat arbitrary because of differences between soils in regard to which pore size corresponds to the limit at which gravity drainage occurs, and the limit at which plants can extract water.

† pF (now obsolete) = log soil moisture tension in cmH_2O.

(from Greenland, 1977)

Weather conditions also play an important role in soil structural development. Exposure of the surface soil to wetting and drying, and to freeze and thaw cycles, contributes to the breakdown of large clods formed in the heavier clay soils. Some soils such as the clay and silty loams are *self-mulching* in the sense that when exposed they tend to break up readily into loose surface aggregates so that relatively less cultivation is necessary to produce a good seed-bed.

However, much that has been published about the agricultural properties of clay soils refers specifically to cool-temperate parts of the world. Young (1974) points out that clay soils in the Tropics often show marked contrasts in mineral composition and associated physical and chemical properties. Despite a clay content which may be as high as 85 per cent, tropical soils are often friable or even floury in condition, being aggregated into very fine structural units. This type of *micro-aggregation* is caused by the cementing properties of free iron oxides, a high content of which is a characteristic of many tropical soils. The result is that they behave more

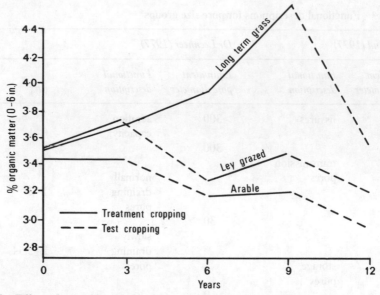

Fig. 4.3 Effect of grazed leys and arable cropping on soil organic matter: Rosamund Experimental Husbandry Farm (from Eagle, 1975)

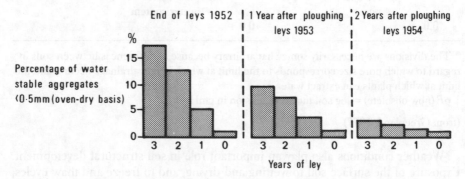

Fig. 4.4 Effects of leys of up to 3 years duration and arable crops on the percentage of water-stable aggregates (<0.5 mm diameter) (from Low, 1975)

like light than heavy soils. Freely permeable and easily penetrated by crop roots, they have, however, a low nutrient-holding capacity. They are, as a result, not only nutrient-deficient but also very susceptible to nutrient loss by leaching. Hence the nutrient-holding properties of the organic matter are proportionately more important in tropical than in temperate soils. However, because of rapid rates of decomposition, organic matter content is generally lower in tropical soils and more highly concentrated near the surface than in temperate soils. Also, under agricultural use it declines at a much greater rate in the former than in the latter.

WORKABILITY

Tillage aims to create the optimum physical environment for crop establishment and production. Its success in this respect will be dependent on the adaptation of the methods used and, more particularly, of the type and timing of management to the particular soil conditions. The latter determine the relative ease with which a soil can be worked and its reaction to cultivation. The terms light, medium and heavy are also indicative of the amount of effort or energy required for tillage (see Table 4.5).

Table 4.5 Force (draw-bar pull) required to draw a plough through soil in relation to soil heaviness as reflected by percentage clay content

Clay content (%)	23.6	30.0	31.1	34.3
Draw-bar pull (lb)	1280	1400	1500	1550
(kg)	580	635	680	703

(from Brown, 1974)

Soil workability is dependent on what is termed its *consistence* – that is, the combination of soil properties material that determine its resistance to externally applied forces such as digging, ploughing or the surface weight and traction of animals and/or vehicles used in cultivation (Brady, 1974). The amount of movement or deformation that will take place is related to the *shear force* which builds up in the soil (see Fig. 4.5). This force results from the friction and/or the cohesion of the

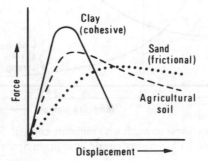

Fig. 4.5 The relationship between force (or draught) and displacement slip (from Davies *et al.*, 1982)

constituent particles. Friction, built up between the implement and the soil material, is a function of the weight, roughness and shape of the constituent particles; and in loose dry coarse soils frictional forces predominate. As illustrated in Fig. 4.6, cohesion is a function of soil moisture content and texture; and the greater the clay content the more important will be the cohesive force.

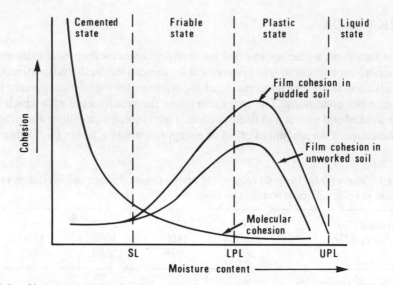

Fig. 4.6a Variation in soil cohesion with moisture content.

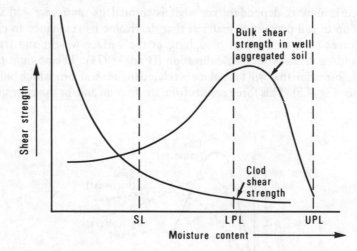

Fig. 4.6b Variation in soil shear strength with moisture content. SL = shrinkage limit; LPL = lower plastic limit; UPL = upper plastic limit (from Spoor, 1975)

Soil consistence is described in terms of the soil condition at different moisture states, e.g. dry and hard; loose and friable; moist and plastic; wet and viscous. Consistency limits, related to critical soil moisture levels, were introduced by Attenberg in 1911 and were originally used by engineers to assess the reaction of soils to the weight of mechanical structures such as buildings and roads. These states and limits (see Table 4.6) refer to medium- and fine-textured soils and are described in the following terms:

Consistency state: cemented or hard; friable; plastic; liquid

Consistency limit: (dry) shrinking; (moist) plastic; (wet) liquid

Coarse-textured soils, with less than 10 per cent clay content, are non-plastic and do not exhibit well-defined consistency limits.

The *shrinkage limit* is that state at which the soil passes from having a moist to having a dry appearance, i.e. the moisture content is just sufficient to fill the pores available at minimum volume as a result of drying. At the *plastic limit* each particle is surrounded by a thin film of water just sufficient to act as a lubricant, while at the *liquid limit* the film of water is thick enough to reduce cohesion between the soil particles and result in 'soil flow' under externally applied forces. Between the shrinkage and plastic limits the soil is friable and moist, a physical condition which allows it to be worked without undue effort and with minimum risk of structural damage. At the plastic limit the risk of *poaching* (puddling) , *smearing* and general structural deterioration increases (see Fig. 4.7). A further distinction is often made between the *lower plastic limit* (or minimum moisture content at which soil puddling occurs and the maximum moisture content at which it can maintain a friable state), and the *upper plastic limit* (at which point the soil changes from a plastic to a viscous or liquid state). The lower limit tends to be reached just before *field capacity* is attained, i.e. at what is described by field surveyors as the *sticky point*.

In medium- and fine-textured soils the *friability index* (i.e. the difference between the shrinkage and plastic limits) is a measure of the acceptable moisture range for the cultivation of many types of crops. The higher the index, the longer will be the period when cultivation is feasible and soil conditions are optimal for tillage. A low index, however, indicates that the soil will change from wet to dry too rapidly for successful cultivation. The *plasticity index* (the difference between the moisture content at the plastic and liquid limits) is often used as a general index of workability in terms of soil susceptibility to structural damage. It is also an indication of the draught or motive force needed for a particular operation. As the index increases, so soil strength or consistence increases logarithmically and a proportionally greater effort is required to attain the desired tilth. Another measure of workability is the ratio between the *plastic limit* and *field capacity*. If the plastic limit is reached before field capacity, soil workability will be very poor indeed (see Table 4.7).

The quantitative material value of these limits and indices depends on the surface area of the soil material as determined by texture and, more particularly, by the clay fraction and the type of clay mineral of which it is composed. Less is known about effect of organic matter, though an increase is known to raise moisture content at the plastic limit. Heavy soils in which the clay fraction is over 28 per cent present the most difficult management problems for arable cultivation. Highly water-retentive, their permeability is low and field drainage is slow. Hence drying out is retarded. Heavy clay soils become plastic when too moist and hard and cloddy when too dry (see Table 4.7). The moisture range within which they can be cultivated is very narrow and field drainage in Britain is usually a necessity. The number of days on which they can be easily worked is small compared with that for other soils (see Table 4.8). *Timeliness* and *speed of cultivation* are therefore of greater importance on these than on any other types of soil.

Consequences of tilling heavy soils beyond their critical consistency limits are

Table 4.6 Moisture content at the plastic limit and at field capacity

Soil texture (surface horizon)	Moisture content (%)		Clay, 0.002 mm (%)	Organic matter (%)
	Plastic limit	Field capacity		
Sandy loam	19.7	26.9	17.0	2.8
Clay loam	35.9	42.0	39.0	6.4
Clay loam	48.8	55.2	58.0	8.8

(from Wilkinson, 1975)

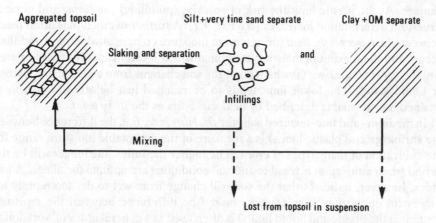

Fig. 4.7 Diagrammatic representation of aggregate breakdown and separation of particle sizes. OM = organic matter (from Davies, 1975)

Table 4.7 Effect of soil moisture on the consistence and workability of heavy soils

Soil moisture	dry	moist	wet	saturated
Consistence	hard, harsh	friable, soft	plastic, sticky	liquid
Bearing strength	high	fairly high	low	very low	practically none
Ease and result of tillage	hard; clods and dust	low-draught soil; crumbles; optimum working conditions	high-draught soil; puddles and compacts	draught lower but soil slakes and ruts	runs together and traction impossible

(from Wilkinson, 1975)

Table 4.8 Influence of rainfall and soil texture on the average number of work days

	Number of work days*								
	February			March			April		
	Light soil	Medium soil	Heavy soil	Light soil	Medium soil	Heavy soil	Light soil	Medium soil	Heavy soil
Wetter than average	3	2	0	16	14	9	21	19	16
Average rainfall	8	5	3	25	24	20	26	23	16
Drier than average	11	9	8	29	29	27	28	26	25

* Work day = day on which cultivation would be satisfactory.

(from Davies *et al.*, 1982)

compaction, which reduces pore spaces and increases bulk density, and *puddling* and *smearing*, which disrupt the continuity of the pore spaces. *Puddling* (or poaching) is a process whereby surface soil aggregates are broken down. It can be caused by either the impact of heavy rain or trampling by animals on a bare or semi-bare surface, or by tillage when the soil is too wet. Some of the clay is dispersed, forming a thin layer over the surface and blocking transmission pores. As a result puddles of water collect in which fine particles settle out and eventually, on drying, form a *crust* or *cap* which may impede seedling emergence. *Smearing* involves the localized spreading and smoothing of soil under the pressure of wheels and ploughshares. Decrease in the volume and size of the pore spaces not only retards water movement but can lead to anaerobic conditions, the production of toxic gases due to incomplete decomposition of organic matter, and to denitrification.

Heavy soils impose serious limitations on cultivation, which are difficult to ameliorate. Flexibility of use is restricted. Roots and winter-harvested crops may be excluded and grass, either on a permanent or rotation basis, has long been the traditional crop of such heavy soils in Britain. However, on intensively managed grassland, where high stock densities are combined with high inputs of nitrogen fertilizer, shallow-rooted grass and low organic-matter levels can give conditions susceptible to *poaching* by animals.

In contrast, the main factor limiting the range and yield of crops on light soils is the tendency to drought during the growing season and, more particularly, during a critical growth phase. A poor nutrient-holding capacity gives rise to more problems than on other soil types. However, the majority of light- and medium-textured soils have fewer inherent limitations than the heavy soils and are often suitable for a wide range of crops. Nevertheless, both fine sandy and silty soils tend to have weakly developed, unstable structures and they are, as a result, even more susceptible to capping than some of the heavier soils. On the sandier soils tillage can result in poor compaction before sowing, giving a loose, 'puffy' and easily blown surface soil. On the other hand, the finer loams and silts tend to compact easily under highly mechanized systems of cultivation.

TRAFFICABILITY

Soil consistence also affects the *carrying capacity*, i.e. the bearing strength or what is more usually termed the *trafficability* of the soil. This is a measure of the amount of movement by wheeled vehicles and/or animals that the soil can carry with minimum draught and soil movement. Light sandy soils have their optimum trafficability under moist conditions. In a dry or wet state, soil movement is excessive and shear strength high, so that heavier or ballasted machinery with greater energy inputs is required. In contrast, on the heavier soils trafficability is highest when they are dry and naturally hard – a condition particularly suitable for combine harvesting for instance. With increasing moisture, compaction increases up to the plastic limit, when trafficability decreases and the danger of surface damage increases rapidly. Surface compaction occurs on all soil types but is found particularly on sand and silt, which pack easily, and on heavy soils when too wet. In the latter case the weight per unit area that can be carried decreases and recourse may need to be made to lighter machines or those with wheels and tyres which distribute the load more widely.

Trafficability and workability are closely related in tillage operations. In the non-tillage operations, particularly those which must be fitted into the beginning or end of the growing season, when soil moisture conditions are becoming critical, trafficability alone may determine ease of accessibility and number of days when conditions are suitable for work with tractors etc.

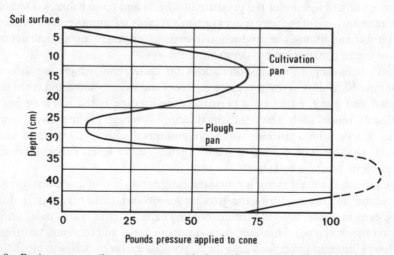

Fig. 4.8 Resistance to soil penetrometer with depth from surface of an agricultural soil showing two compacted zones, i.e. a cultivation pan and a plough pan (from Harrod, 1975)

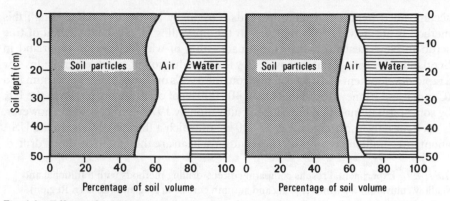

Fig. 4.9 Effects of cultivations and animal stocking on the volume of soil occupied by air, water and solids (from Simpson, 1980)

OVERCULTIVATION

Spoor (1975) has made a distinction between 'positive' and 'negative' cultivation, the latter resulting in unwanted but often unavoidable soil deformation arising from non-tillage operations or from tractor and plough use during positive cultivation. Problems arising from increasing soil compaction either by horses or tractors, or by frequent ploughing to a constant depth, were early recognized and by the late nineteenth century subsoil tines were being used to break up *plough pans* (see Figs 4.8 and 4.9). Increasing intensity of arable farming in recent decades has exacerbated the problem. Impact has been increased with increasing size and, particularly, specialization of implements for cultivation. It is not unusual for a field to be tracked a dozen times (exclusive of ploughing) during the growing season (Simpson, 1980). In extreme cases, over 200 passes per field by wheeled vehicles during the production of some horticultural crops has been recorded. 'Bed' or 'tramline' systems in which traffic follows parallel tracks where no crops are grown are being adopted to minimize surface compaction. Cereal monoculture and increasingly high inputs of inorganic fertilizers have resulted in the decline of soil organic matter with a consequent deterioration of soil structure. Light soils exposed by cultivation have become more prone to wind erosion; silty and heavier soils to puddling and capping after heavy rain. Another result of surface compaction is the formation of large clods of varying strength and resistance to breakdown.

MINIMUM OR ZERO TILLAGE

With the use of herbicides and the burning of straw and trash, tillage aims and methods in intensive farming systems have recently undergone radical changes. The depth of tillage has been reduced and the rotary cultivator has replaced the plough as the primary tillage implement, particularly on heavy land. In addition, there has been an increasing trend to reduce the number of cultivations in order to avoid the undesirable results of overcultivation. Reduction or elimination of tillage has been accompanied by direct drilling of seeds. Originally known as 'sod-seeding', this method is not new. It was employed in the upgrading of hill-land cleared of tree scrub in New Zealand and in the broadcasting of white clover on moorland in Scotland. The opportunites provided by the introduction of suitable herbicides saw a rapid development of direct drilling from the 1950s onwards. It has been estimated (Cannell and Finney, 1973) that about 400 000 ha of cereals were grown in the UK by some form of modified or reduced cultivation by 1970. Direct drilling, however, is used for only a small area, e.g. 55 000 ha, of which a third is cereals. In the USA about three million hectares of wheat and soya bean are thought to be direct drilled.

Table 4.9 Experimental results comparing direct-drilling methods with traditional and shallow cultivation for winter, spring and autumn cereals in England (Eastern Region)

	Traditional cultivation (mould-board plough)	Shallow cultivation (tine or disc plough)	Direct drilling
Percentage autumn cereal treated	65	20	15
Man-hours ha^{-1} (56 kW tractor)	5.5	3.25	1
Mean yield (tonne ha^{-1})			
Winter wheat (all soils)	6.2	6.2	6.3
Winter barley (light soil)	5.8	5.6	5.5
Winter barley (heavy soil)	6.3	6.1	6.2
Spring cereals (heavy soil)	7.0	4.1	3.8

(from MAFF, 1983)

Direct drilling has been extensively researched and despite continuing debate it would seem that it can have advantages and maintain adequate soil conditions without traditional soil disturbance (see Table 4.9). Its greatest success and potential would appear to be on heavy soils, where timeliness of cultivation is so critical. It has been used on a large scale on livestock farms for drilling kale, turnips, swedes and oil-seed rape. The advantages under favourable soil conditions are a high plant population for the same rate of seed application; earlier germination and quicker crop establishment; and more even plant distribution. The main disadvantages are that direct drilling favours perennial weeds and certain pests (including slugs) and diseases, while rusts and mildew can overwinter on volunteer cereals (i.e., plants which grow from seeds or shoots of harvested crop). Also, nutrient leaching may be greater than on ploughed land. It has been suggested that direct-drilled crops often require more nitrogen fertilizer to reach maximum yield than tilled ones either because of greater weed competition or because of slower crop establishment. Soils are inevitably more compacted after direct drilling; higher density and greater mechanical strength together with a more level surface, allow easier trafficability.

Porosity and, in particular, the proportion of large pores decrease (see Table 4.10). The pore spaces, however, may often be more continuous, partly as a result of an increase in numbers of earthworms, so that water infiltration is improved. Compaction can also be accompanied by increased soil moisture on sandy loams and silts; on finer textured soils, however, it has been found that surface waterlogging may occur. In some cases compaction can be accompanied by a greater range in soil temperature, in others by an increased thermal capacity.

Table 4.10   Percentage total porosity and pore-size distribution by volume

Total pore space		Coarse pores greater than 30 µ		Fine pores less than 30 µ	
DD	Cul.	DD	Cul.	DD	Cul.
43	48	10	17	33	31

DD = direct-drilled; Cul. = deep cultivated.

(from Davies *et al.*, 1982)

Table 4.11 illustrates the possible effects of ploughing and direct drilling on soil structure and content of organic matter on a site which was previously permanent pasture. Decline in both aggregation and organic matter takes place with ploughing and direct drilling. However, the reduction in both is more normal with direct drilling particularly in the top 5 cm of the soil.

Less is known about the effects on root growth and crop performance of soil changes resulting from direct drilling. While root elongation varies with soil strength, the limiting bulk density can vary widely between different soils; and the restrictive effect of compaction may be mitigated by the channels left by dead and decaying roots and earthworms. In the case of tap-rooted crops (e.g. cotton, kale, sugar beet etc.) shallow-rooted systems tend to develop. In the case of cereals, root growth may initially be restricted after direct drilling though later growth and yield are unaffected. However, there is still a need to examine the long-term consequences of direct drilling on root growth. With improved drills and herbicides the modified 'soil physical environment may become the principal limitation and there is a need to define conditions where, with proper management, the technique can be safely used' (Cannell and Finney, 1973, 189).

DRAINAGE

While the importance of tillage for crop establishment and performance may, under certain circumstances, have declined, the management and regulation of soil moisture has retained it role as one of the most vital of the cultivation processes.

In many parts of the world a soil water surplus for part or whole of the year limits the amount of time available for crop growth. Some form of soil drainage is thus

Table 4.11 Stability of soil aggregates and organic-matter content in a sandy loam after ten years ploughing or direct drilling

Depth (cm)	Percentage of aggregates in each stability category*		Organic matter content (%)
	Low stability	High stability	
Ploughed plots			
0–2.5	46	7	7.8
2.5–5	58	2	7.4
5–10	30	10	7.7
10–15	20	20	7.7
15–20	39	10	6.1
Direct-drilled plots			
0–2.5	20	36	9.5
2.5–5	26	6	7.8
5–10	25	17	7.4
10–15	22	11	7.4
15–20	22	21	6.1
Grass reference area			
0–2.5	6	56	10.0
2.5–5	5	26	7.4
5–10	0	37	7.2
10–15	11	20	7.2
15–20	31	29	5.8

* Each result is the mean of about 200 soil aggregates. Low-stability aggregates required <25 droplets of water for disintegration; high-stability aggregates required >200 droplets.

(from Cannell and Finney, 1973)

necessary for successful cultivation. The basic aim of all drainage systems is to remove excess soil water and thereby allow better aeration and deeper and more extensive root development. According to Thomasson (1975, 7) the goal in England is 'to imitate, as far as possible, the chalkland situation, ensuring that free water has been removed and the soil returns to fields capacity within forty-eight hours'. It has been estimated that over half the farmland of England requires artificial drainage to remove excess water. Among the conditions which necessitate drainage are: a permanently or seasonally high water-table; high rainfall and low evaporation rates combined with heavy soils; an impermeable subsoil; or acid peats with a particularly high water-holding capacity. While a well-drained soil may return to field capacity after rain in 48h, a poorly drained one may take 7–10 days.

Land drainage has a long history dating, at least in Britain, from Roman times. However, it was not until the seventeenth century that more advanced techniques began to evolve and the extent of areas improved or reclaimed increased dramatically. The effectiveness and longevity of a particular type of drainage depend essentially on the design of the system, the nature of the soil and the quality of management.

The preferred design of a drainage system in terms of the depth, spacing and permeability of the evacuation channels will be that best able to cope with the soil hydrological conditions and, at the same time, give an economic return on the capital invested in its development and maintenance. Design will also be dependent on the type of crop to be grown, its susceptibility to excess soil water, cultivation requirements, the time and type of harvesting and, not least, its economic value. The essential requirement of the system is usually to prevent waterlogging within the top 50 cm of the soil in all but exceptional circumstances. Normally the depth to which ditches are cut or drains laid varies between 70 and 120 cm, but a minimum of 50 cm and a maximum of 2.0 m are possible. For under-drainage, tile drains or, more commonly now, plastic pipes (with or without a permeable fill above them) are used. The former are about 25 cm long and 10–15 cm in diameter, the latter can be put up to 150 m in length. Spacing, which affects the rate at which the water-table rises and falls, ranges from 5 to 40 m, with an average of about 10 m being normal. Originally adapted to one-way ploughing, drain spacing tends often to be more a function of local tradition than the hydraulic properties of the soil. On heavier land in the West of England and Wales and in Scotland closer spacing has long been the norm.

The need for artificial drainage and its efficiency are related to the size, shape and distribution of the pore spaces, which determine the hydraulic conductivity of the soil. In clay soils the conductivity is less than 0.1 mm day^{-1} and nearly all effective water movement is confined to the A-horizon. Drainage will, under these circumstances, be ineffective unless the conductivity of the subsoil is ameliorated. The methods presently used are those which physically shatter and lift the subsoil, thereby increasing the number and size of cracks and fissures. This is effected either by subsoiling and/or mole-draining.

Moling involves the formation of a subsurface unlined circular channel (500–700 cm depth). A 'bullet' *c.* 75 mm in diameter, mounted on the end of a blade (2 cm thick; 30 cm wide) is drawn through the soil; and a spherical plug 10 cm in diameter is towed behind to enlarge and smooth the channel. Moling is usually only successful in stone-free heavy soils which do not slake in water. *Subsoiling* (or square-moling) covers a wide range of operations designed to shatter the subsoil at a depth of 450–600 cm, particularly compacted horizons and/or *plough pans*. It differs from moling in the absence of a defined channel and in the greater intensity of shattering achieved. Subsoiling produces vertical or subvertical fissures, which improve soil permeability and increase both vertical and horizontal water movement to the drains. The final success, however, of any of these systems depends on there being a sufficient 'head' or gradient to carry water away; otherwise pumping is necessary.

The period during which the soil benefits from an effective drainage system is that when the soil is above field capacity. In Britain the *drainage winter* commences when drains begin to run; and the drainage climatic index (i.e. millimetres of excess winter rain divided by the duration in days of the winter drainage period) is a measure of the rainfall per day that needs to be evacuated (MAFF, 1976b). Design of normal drainage is often based on probabilities of 7 or 10 mm day^{-1}, and that of mole-drainage, of 1 mm day^{-1}, without, however, taking the maximum upper limits possible into consideration. However, in high-rainfall areas, where it is often impossible to identify precisely the date on which the drains start and stop running and where the soil may rarely fall below field capacity for any length of time, it is not possible to make such estimates of drainage requirements. Above a certain level of rainfall (*c.* 130 cm, UK) there is a definite limit in the improvement that can be achieved. Conversely, the *drainage summer* is the period when drains stop running

Table 4.12 Mean work days and dates of return to field capacity

Date of return to field capacity	Drainage categories			
	Good	*Moderate*	*Poor*	*Bad*
Early-Sept.	5	5	4	3
Mid-Sept.	6	5	4	3
Late-Sept.	13	12	10	8
Early-Oct.	16	14	12	10
Mid-Oct.	25	22	19	16
Late-Oct.	26	23	20	17
Early-Nov.	31	27	24	21
Mid-Nov.	36	32	28	25
Late-Nov.	35	29	25	21
Early-Dec.	45	40	36	33
Mid-Dec.	48	45	41	38
Late-Dec.	59	53	48	44

(from Smith, 1975)

and the soil returns to or falls below field capacity. Then the soil should be dry enough for drainage systems to be laid successfully with the minimum of soil damage. It has been estimated that a soil water deficit of 50 cm is needed for moling, and of 100 mm for subsoiling. This is also a measure of the *accessibility period* for optimum cultivation (see Table 4.12). Among the benefits of a well-drained soil are timeliness of cultivation and easier soil management as well as increased crop yields.

5

Nutrient cycling

The soil is the nutrient pool or reservoir from which, in both unmanaged and all but a few specialized agricultural systems, the nutrients essential for plant growth and development are drawn. Varying amounts of some fourteen elements are required. They include the *macronutrients* needed in relatively large and the *micronutrients* (or trace elements) in very small quantities. Of the former, the most important are nitrogen, phosphorus and potassium – the 'Big Three' – the primary fertilizer elements used in agriculture. The latter comprise iron, manganese, boron, molybdenum, copper, zinc, chlorine and cobalt and, with the exception of iron and manganese, occur naturally in small amounts in the soil.

Table 5.1 Principal ionic forms of nutrients absorbed by crops

Element	*Cations*	*Anions*
Macronutrients		
Nitrogen	$(NH_4)^+$ (ammonium)	$(NO_3)^-$ (nitrate)
Calcium	Ca^{2+}	–
Magnesium	Mg^{2+}	–
Potassium	K^+	–
Phosphorus	–	$(HPO_4)^{2-}$ ⎫ (phosphates)
	–	$(H_2PO_4)^-$ ⎭
Sulphur	–	$(SO_4)^{2-}$ (sulphate)
Micronutrients		
Copper	Cu^{2+}	–
Iron	Fe^{2+}	–
Manganese	Mn^{2+}	–
Zinc	Zn^{2+}	–
Boron	–	$(BO_3)^{3-}$
Molybdenum	–	$(MoO_4)^{2-}$
Chlorine	–	Cl^-

Nutrient uptake, however, is dependent on the mineral element being *available* – i.e. in a chemical form that can be readily adsorbed by the plant root system – at or near the root–soil solution interface (Hillel, 1972). The available elements are those present in relatively simple soluble forms, either as anions in the soil solution or as cations capable of being adsorbed, and hence retained, on the surface of the clay–humus colloidal complex (see Table 5.1). The amount of available nutrients and their ease of adsorption is also closely dependent on the physical as well as the chemical condition of the soil. On the one hand, the amount of available nutrients retained by the soil is determined by its texture and structure but, more particularly, by the *cation exchange capacity* of the colloidal complex. On the other hand, uptake can be affected by the ease with which the crop root system can develop laterally and vertically to exploit the soil. In addition, less soluble and hence less readily available nutrients occur in the soil in complex organic or inorganic compounds from which they may eventually be gradually released either by decomposition or by rock weathering as the case may be. In uncultivated soils most of the nitrogen, phosphorus and sulphur is held in the form of organic compounds and the proportion readily available (see Table 5.2) is relatively small. In contrast, calcium, magnesium and potassium occur in mineral compounds. The first two are easily released in an available soluble form by rock weathering. Potassium is less readily available.

Table 5.2   Availability and mobility of principal crop nutrients in soil

	Soil supply			
Macronutrients	Amount	Availability	Crop removal	Leaching
Nitrogen	high	low	high	low to high
Phosphorus	high	low	low	low
Sulphur	high	low	high	high
Potassium	high (except in sandy soils)	moderate	high ('luxury consumption')	low to high
Calcium and magnesium	high	high	low	high

In unmanaged ecosystems nutrients taken up by plants are eventually recycled, i.e. returned to the soil by the process of organic decomposition. The size and composition of the pool of soil nutrients and the rate at which they move out of and back into the soil varies from one ecosystem to another. However, the accepted concept of the *biological cycle* assumes the maintenance of a balance between nutrient output and input, and hence of the size of the nutrient pool itself, once the system has attained a 'steady-state'. Losses of nutrient by soil erosion, leaching, volatilization or by export of living or dead organic matter are considered, under these circumstances, to be small. Further, they are assumed to be adequately compensated for by inputs of nutrients via precipitation, by biological fixation of gaseous nitrogen and by imports of either mineral and/or organic material from other systems (see Fig. 5.1).

In contrast, the steady-state or balanced nutrient cycle is neither characteristic nor easy to achieve in agro-ecosystems. The volume and rate of the nutrient cycle in agro-ecosystems vary according to the type and intensity of farming practised. The cycle is also more open and extensive, with greater inputs and/or outputs than in the more localized nutrient cycle in unmanaged ecosystems. In some agro-ecosystems a near balance may be achieved; in others the losses are greater than the gains; while in most modern intensive systems of farming the aim has been increased agricultural yield with the need for even greater inputs of nutrients. In general the amount of nutrients taken up by crops is, particularly for nitrogen and potassium, greater than by similar uncultivated plant life-forms. However, both take up varying amounts of the essential nutrients. Nutrient 'demand' varies not only with the type of crop and its yield potential but also with the nature of the harvested part and the length of the growing season available for its production. Temperate cereals, for instance, take up less nutrients per unit area than 'green' or root crops; the latter can remove up to five times as many nutrients as cereals from the same land (Cooke, 1972). High-yielding C4 crops of warm climates, such as sugar cane and maize, are much more demanding. Also, within one type of crop there may exist numerous varieties (*cultivars*) and strains which differ in their yield potential and hence in their nutrient requirements.

Losses of nutrient from the agro-ecosystem as a result of crop and/or livestock removal inevitably constitute a continued and heavy drain on the soil pool of available nutrients. On the whole the loss of nutrients in livestock or livestock products sold off a farm is less than from arable cash crops, particularly when animal excrement is returned to the land from which the feedstuffs were grown. However, in all agro-ecosystems losses have to be made good by management which aims to supply, in one form or another, the correct volume of nutrients required to maintain or increase productivity. Supply can either be from within and/or beyond the farm, in organic and/or inorganic form, depending on the type of agro-ecosystem.

ORGANIC NUTRIENT SUPPLY

In the uncultivated or otherwise unfertilized ecosystem, organic matter is the principal nutrient reservoir within the soil. It consists of living roots, soil animals, micro-organisms, and dead organic matter (mainly plant-derived) in various stages of decomposition. The volume and vertical distribution of the last will be dependent on the amount and nutritive value of the original organic material contributed on or below the soil surface. Organic matter with a low carbon to nitrogen (C:N) ratio can support a large population of soil micro-organisms, among which the most important are the bacterial decomposers, and earthworms. The latter play a vital role in the initial comminution and eventual distribution of organic matter through the soil mineral material. Organic matter in agricultural soils usually has a C:N ratio of *c.* 1:12 as compared to 1:20 in unmanaged ecosystems.

Decomposition results in an initial relatively rapid release by *mineralization* of

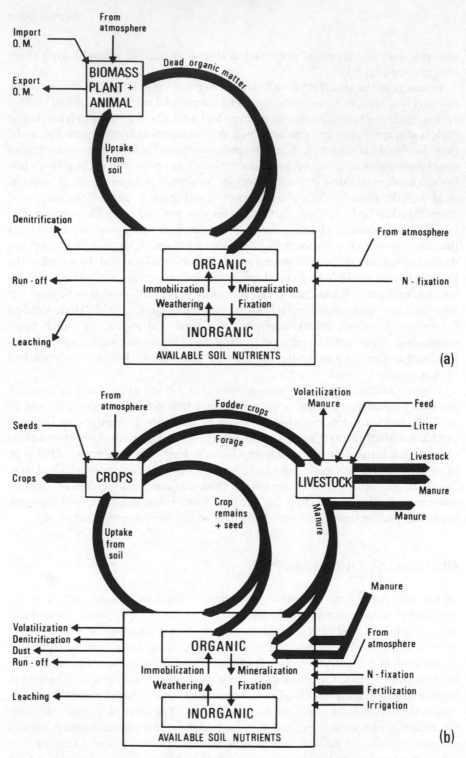

Fig. 5.1 Nutrient cycling in an unmanaged ecosystem (a) and an agro-ecosystem (b). OM = organic matter (from Tivy, 1987)

some of the constituent nutrients in an available form. The remainder are retained, immobilized for a longer period of time in the slowly *humifying* organic matter, the end-product being a relatively stable persistent *colloidal humus* fraction. This becomes bound to the fine mineral fraction (less than 0.001 mm) to form what is known as the soil *colloidal complex*. It is considered that true humic colloids can only be formed in the presence of clay particles on whose surface they are adsorbed or, at least, that they are long lived only when so adsorbed. Hence it is difficult to maintain an appreciable humus content in dry sandy soils (Russell, 1973).

Humus or organic matter in the general sense (usually expressed as the percentage of organic carbon) makes an important contribution both directly and indirectly to soil fertility. With a high cation exchange capacity it increases the nutritive-holding properties of the soil and helps to counteract nutrient loss by leaching, particularly from freely drained, coarse-textured and acid soils. The presence of humus also increases the availability of phosphorus in soils with an appreciable amount of iron and aluminium. It can also form organometallic compounds with micronutrients such as iron and copper and hence check their loss by leaching. In addition, organic matter improves the physical condition of the soil and facilitates management by increasing water retention and mitigating the 'droughtiness' of light soils; and by reducing bulk density and improving aeration and drainage in heavy soils. Its presence is essential for the formation of stable, but porous, soil aggregates associated with a crumb structure.

The amount of organic matter in a soil depends on the rate at which it is added, on the one hand, and the rate at which it is oxidized, on the other. Cultivation affects both processes in such a way as to reduce the organic matter content. The original source of natural vegetation is removed. Tillage opens up the soil and, by increasing temperature and aeration, speeds up the rate of decomposition – by as much as four times in soils in the humid Tropics (Sanchez and Salinas 1981). As indicated in Fig 5.2 the organic matter content in soils cleared of tropical forest or savanna vegetation drops by half in the top 15 cm. Annual rates of decomposition are highest in bare fallows and continuous cultivation. In the temperate localities, which have been under cultivation for a longer period, rates are comparatively very slow, though those in land under continuous cultivation are over double those in land under crop rotation systems. Loss of nitrogen by volatilization and denitrification and of nitrogen and other nutrients released by leaching in drainage water is accelerated. A surface 'mulch' of organic matter applied during fallow periods when the soil might otherwise lie bare can serve to lower temperatures sufficiently to reduce the rate of decomposition and, in addition, provide protection against erosion.

Addition of organic matter is the traditional method in many farming systems of making good nutrient loss by cropping and leaching. Crops like cereals and grass leave an appreciable residue of roots, stalks and leaves which can eventually be ploughed in; or a green crop (*green manure*) may be grown for this purpose. Crop residues alone, however, can replace only a very small fraction of the organic matter lost as a result of cultivation. The most commonly used and effective means of maintaining soil organic levels and ensuring nutrient return is to sow land down to grass for a longer or shorter period of time. Such rotation grass (or 'ley') usually consists of mixtures of agricultural grasses and legumes of which clover, vetch and

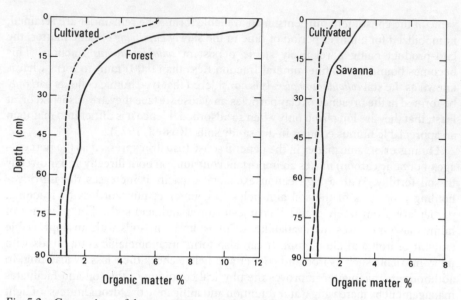

Fig. 5.2 Comparison of the percentage organic matter in tropical forest and savanna soils before and after cultivation (from Sanchez, 1976)

lucerne (alfalfa) are the most commonly used. The dense network of grass roots builds up organic matter in the soil (see Table 5.3) as well as aiding the development of a crumb structure. The legumes are of particular agronomic importance because of the symbiotic association of nitrogen-fixing bacteria (*Rhizobium* spp.) in their root nodules. It is estimated that a good legume crop will increase the nitrogen content in the soil by over 100–200 kg per hectare per year in temperate regions (see Table 5.4). There is, however, a paucity of data about the amount of nitrogen fixed in farming systems other than in experimental plots. In addition, green-manures composed of such legumes as peas and beans are also high nitrogen-fixers. By and large, however, the cash or food value of these crops usually outweighs that of their nitrogen-fixing properties.

Experimental work has indicated that it is difficult, in practice, for traditional farming methods (based on ploughing to 15–20 cm depth and with adequate fertilizer input) to vary the organic content of the soil by much more than 0.1–0.2 per cent over 20 years *without* the use of long leys and frequent application of farmyard manure. The latter is still a commonly used method not only of maintaining organic content (see Table 5.5) but also of maintaining the nutrient balance in a wide range of farming systems.

Organic fertilizers or *manures* comprise wastes and residues from crops and livestock. In comparison to inorganic mineral fertilizer, they are relatively high in carbon but low in nutrients. However, they were before the Second World War the main source of nutrients for the majority of farms in the now-developed countries of the world and they still retain this role in the more traditional systems of farming in these and in developing countries. *Farmyard manure* (FYM) was originally the most important organic source of nutrients in terms of the quantity used. Normally

Table 5.3   Effect of leys on percentage of soil organic carbon*

Types of system	Percentage of soil organic carbon					
	Old arable (age in years)			Old pasture (age in years)		
	6	12	18	6	12	18
Old pasture	–	–	–	3.17	3.74	3.36
New pasture	1.75	2.10	–	2.97	3.36	–
All arable	1.47	1.34	1.46	2.67	2.14	2.08
Leys 3 years arable after 3 years of:						
grass/clover	1.55	1.48	1.61	2.66	2.35	2.22
grass with nitrogen	1.50	1.44	1.62	2.60	2.27	2.18
lucerne	1.51	1.34	1.54	2.62	2.21	2.06

* Rothamsted: six-course rotation; sampling 0.22 cm.

(from Russell, 1977)

Table 5.4   Estimated fixation of nitrogen by nodulated crops in temperate and tropical regions

Temperate		Tropical and subtropical	
Crop	Nitrogen fixation ($kg\ ha^{-1}\ yr^{-1}$)	Crop	Nitrogen fixation ($kg\ ha^{-1}\ yr^{-1}$)
Clovers	55–600	Grazed grass/	
Lucerne	55–400	legume pasture	10–129
Soy	90–200	Beans	64
Pea	33–160	Pigeon pea	97–152
Median	*c.* 200	*Median*	*c.* 100

(from White, 1987)

composed of a mixture of mainly cattle dung and urine with straw, in a solid to liquid ratio of 3:1, its nutrient content is variable, depending on the size, age and condition of the animals from which it is derived (see Table 5.6).

In Third World countries FYM is generally nutrient-deficient as a result of poor animal nutrition. It has a higher C:N ratio, which results in a lower nitrogen release and greater competition between the soil mirco-organisms and the crop, than the richer manures of temperate areas. Of the nitrogen released during organic decomposition, about a third becomes rapidly available for use; the remainder tends to be locked up for much longer in the more resistant humus fraction of the soil (see

Table 5.5 Effect of farmyard manure and ley grass on percentage of soil organic carbon

	Percentage of soil organic carbon			
	Grass/clover	Lucerne	Arable: 1-year ley	Arable: no ley
No farmyard manure	1.30	0.95	0.95	0.88
Plus farmyard manure (35 tonne ha^{-1} annum^{-5})	1.32	1.13	1.04	0.98

(from Russell, 1977)

Table 5.6 Composition of farmyard manures and fresh slurries

	Percentage fresh weight			
Type	approx. dry matter	Nitrogen (N)	Phosphate (P$_2$O$_5$)	Potassium (K$_2$O)
Farmyard-manure				
Cattle	25	0.6	0.3	0.04
Pigs	25	0.6	0.6	0.04
Poultry				
deep litter	70	1.7	1.8	0.40
broiler litter	70	2.4	2.2	0.22
in-house air-dried droppings	70	4.2	2.8	0.40
Slurry (fresh and undiluted)				
Cattle	10	0.5	0.2	0.05
Pigs				
dry meal fed	10	0.6	0.4	0.03
pipeline fed	10–6	0.5	0.2	0.03
whey fed	2–4	0.3	0.2	0.03
Poultry	25	1.4	1.1	0.12

(from Archer, 1985)

Fig. 5.5). FYM alone is not really capable of returning more than about 50% of the nitrogen, phosphorus and potassium removed in the crops fed to animals. The straw component of FYM is difficult to use on its own. It has a very high C:N ratio and decomposition requires a supplementary source of nitrogen. Soil organisms (bacteria and fungi) can compete successfully with the crop for nitrogen and hence temporarily check plant growth rates. It can also aggravate soil water-logging in winter and, under resulting anaerobic conditions, organic acids and ethylene gas toxic to plants may be produced. In addition, cereal straw may contain the residues of herbicides which could be detrimental to the succeeding crop.

During the past 35–40 years *liquid organic manures* (or slurries) have increased in importance as a result, mainly, of the evolution of intensive livestock farming systems. Increasing specialization of agriculture has been accompanied by a locational separation of crop and livestock production. Hence areas where straw is produced have become divorced from those where it can be used and prices are not sufficiently high to compensate for transport costs. Also the introduction of high-yielding but *short-stalked* cereal varieties has reduced the amount of straw available. And shortages have been further exacerbated by the practice of stubble-burning, particularly in areas of continuous cereal cultivation with minimum or zero-tillage. While burning obviates the necessity of ploughing in the straw and helps to clean the ground of weeds and disease organisms, it is thought to accelerate nutrient loss. All the nitrogen content tends to be vaporized, while the phosphorus and potassium released in the ashes become extremely susceptible to wind and/or water erosion.

Methods of livestock management have also changed markedly in the same period. Straw-bedding is now less common and hence less FYM is made. Animals, when kept indoors, stand or lie on slatted floors which facilitate the evacuation of excrement. Much livestock excrement is now disposed of as *liquid slurry* (dung plus urine mixed with water) and is applied directly to the land in this form. This practice is not new. It was the basis of the traditional Gülle system characteristic of grassland farming in Switzerland. The collection and storage of urine from livestock in underground tanks was also common in Britain in the nineteenth century (Cooke, 1977); it has been retained in many European countries other than Britain. Nowadays the use of liquid manure, collected in slurry tanks and piped onto fields where it is applied by spraying is characteristic, particularly of large intensive grass-based dairy and pig farms in South West Scotland.

Another organic source of nutrients is sewage sludge, produced by modern sewage works and often referred to as *organic fertilizer*, rather than manure. Sludge containing as much nitrogen and phosphorus as FYM is agriculturally valuable. However, the nutrient content of sewage sludge varies greatly and it is always lower in potassium content than FYM. In addition, urban/industrial waste often contains non-essential metallic elements in quantities that can be toxic to plants and animals. Too long and frequent application of sludge may eventually result in the concentration at toxic levels of such heavy metals as zinc, copper, nickel, cobalt, boron, lead and mercury. And since these become immobile in the soil they cannot readily be removed. Sewage sludge is the modern counterpart of 'night soil', the human and livestock excrement still used in tropical 'garden-culture' and of such crops as rice in many parts of Monsoon Asia.

INORGANIC NUTRIENT SUPPLY

Until the nineteenth century marl and lime (both used in Britain since Roman times) were the only inorganic sources of nutrients available. Deposits of sodium nitrate were the first to be exploited, and the process whereby superphosphate was produced by dissolving bones in sulphuric acid (developed at Rothamsted in 1840) heralded the beginning of the modern chemical fertilizer industry. It was not until the Second World War, however, that scientific developments and demands for increased food production in many western countries stimulated the use of inorganic fertilizers. Their production in a concentrated (dry) form made handling easy and safe and resulted in low transport and application costs.

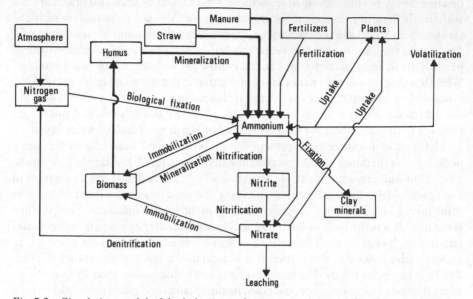

Fig. 5.3   Simulation model of the behaviour of nitrogen in the soil (after Frissel, 1978)

The inorganic fertilizers allowed the farmer to supplement and eventually become independent of organic sources of nutrients. Indeed they now supply more nutrients than the latter in most developed countries. They are usually in the form of simple chemical compounds derived directly from rocks, processed from mineral sources or manufactured from simple elements. The latter are the so-called 'artificial' or more accurately 'synthetic' fertilizers (a term often used erroneously to cover all inorganic nutrients). In most countries the bulk of the inorganic fertilizers comprise nitrogen, phosphorus and potassium, together with calcium and magnesium. The efficiency with which these inorganic nutrients are cycled and the resultant ratio of input to output in products varies. It depends on the particular nutrient, the form in which it is applied as well as its solubility and mobility in the soil and its susceptibility to loss by leaching, drainage, volatilization, denitrification and soil erosion (see Fig. 5.3).

NITROGEN (see Figs. 5.3 and 5.4)

There is now a wide range of nitrogen fertilizers synthetically produced from atmospheric nitrogen. Initially both ammonium salts and nitrates were the most commonly used forms, particularly *ammonium sulphate* (see Table 5.7). The latter

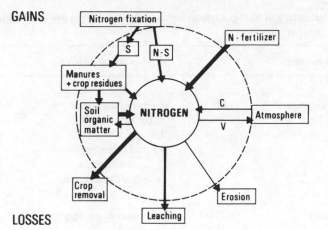

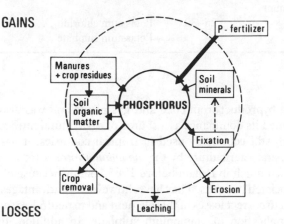

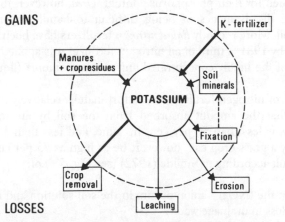

Fig. 5.4 Gains and losses in cycling of nitrogen, phosphorus and potassium in agro-ecosystems S = symbiotic; N–S = non-symbiotic; C = condensation and precipitation; V = volatilization (adapted from Buckman and Brady, 1960)

Table 5.7 Percentage of nitrogen, phosphorus and potassium in low- and high-yielding forms of inorganic fertilizers

Low-nutrient yielding	%	High-nutrient yielding	%
Nitrogen			
Ammonium sulphate	21	Ammonium nitrate (or nitro-chalk)	35
		Urea	46
		Liquid nitrogen	
		aqua ammonia	25–29
		anhydrous ammonia	82
Phosphorus			
Superphosphate	*c.* 20	Ammonium phosphate	*c.* 30
Potassium			
Potassium/magnesium sulphate	25–30	Potassium chloride	50–60
		Potassium sulphate	*c.* 50

(from Tivy, 1987)

was an important byproduct from coke and gas plants and was also synthetically produced; it retained its dominance as a source of agricultural nitrogen for nearly a century. Although still commonly used in tropical countries, it has been all but displaced in temperate agriculture by the *ammonium nitrates* (or *nitro-chalk*). The latter were not used much in Britain before 1965, mainly because of difficulties of handling and associated fire risks. They have, however, the advantages of containing sufficient calcium to correct losses of this element and to neutralize soil acidification caused by the application of ammonium sulphate. In addition, the *ammonium phosphates* are valued for their phosphorus content. *Urea*, however, though high in nitrogen, has the disadvantages of being difficult to handle, hygroscopic and unstable in the soil. More recently *liquid nitrogen* fertilizers have been pioneered in the USA, where, by 1981, a third of all nitrogen fertilizer was applied in this form, particularly that of the highly concentrated anhydrous ammonia (liquid ammonia gas).

The efficiency of nitrogen fertilizers is, unfortunately, relatively low and it has been estimated that the amount absorbed from the soil by an arable crop in temperate regions is less than 50 per cent in some, and less than 30 per cent in others. Uptake by a grass crop can, however, be as high as 75 per cent. This low efficiency is a result according to Smilde (1972) (see Fig. 5.4) of:

1. *Nitrate leaching*: the $(NO_3)^-$ anion occurs in the soil solution and hence is very susceptible to loss in drainage water.
2. *Denitrification*: the bacterial reduction of nitrate to gaseous nitrogen which occurs particularly under conditions of poor soil drainage and aeration.
3. *Volatilization* of ammonia from excreta, animal manure and crop residues on the

soil surface, from the leaves of heavily fertilized plants and from ammonium compounds applied on or near the soil surface.

The problems of reducing nitrogen losses vary with environmental conditions and with farming systems. In humid temperate and tropical regions the main efforts have been directed towards adapting the amount and time of application of nitrogen fertilizers to local soil and climatic conditions in such a way as to maximize uptake by crops and minimize loss by leaching. Loss by volatilization is most serious in semi-arid areas, and in paddy rice cultivation. Under traditional methods of cultivation volatilization losses of up to 60 per cent nitrogen applied can occur, while losses from denitrification, at approximately 10 per cent, are much less important than was formerly considered (Cooke, 1981).

PHOSPHORUS (see Fig. 5.4)

Although the total phosphorus content of the average mineral soil compares favourably with that of nitrogen, it is much lower than either potassium, calcium or magnesium. In addition, nearly all the phosphorus occurs in a form unavailable to plants; hence it is the nutrient most frequently limiting yield on newly farmed and formerly unfertilized soils in, for instance, underdeveloped countries (Russell, 1977). In developed intensively farmed areas phosphorus is used to give young crops an initial boost, to attain yields near the maximum potential of the crop and to sustain the amount of available phosphorus in the soils at levels necessary for the particular farming system.

The main form of the mineral fertilizer is *phosphate* derived from sedimentary rocks found in many parts of the world. Much finely ground phosphatic rock, suitable for direct application, comes from North Africa: for example, *phosphal* (an aluminium–calcium phosphate), which is produced in Senegal by calcining and grinding phosphate (Cooke, 1972). Indeed the first chemically manufactured water-soluble fertilizer was *superphosphate*, produced by treating rock phosphate with sulphuric acid. Together with *basic slag*, it came into use in the mid-nineteenth century and made a revolutionary contribution to increasing agricultural yields. Basic slag, in particular, was important in the successful establishment of wild white clover and the concomitant improved of long leys. A byproduct of iron-ore smelting, it contained, in addition to phosphorus, 25–30 per cent calcium, as well as varying amounts of other minerals; a ton of slag might contain the equivalent of two-thirds of the calcium found in the same amount of good limestone. It was formerly extensively used as a 'top-dressing' to improve and maintain the condition of permanent grassland. However, recent technical innovations in the steel industry have resulted in a rapid decline in the supply of high-grade slag. Indeed both basic slag and superphosphate have now been all but replaced by the more concentrated *ammonium phosphates* which, in addition, are used in the production of other compounds. Today nearly 90 per cent of the phosphate used in the UK is sold as compound fertilizer.

Phosporus (see Fig. 5.4) is the most 'temperamental' of the major crop nutrients, with low mobility and efficiency of use. Less susceptible to leaching than either

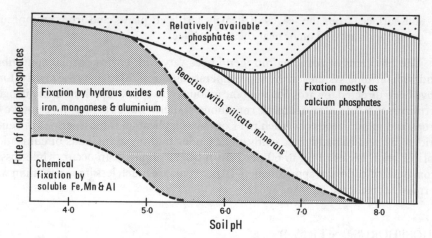

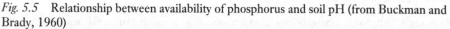

Fig. 5.5 Relationship between availability of phosphorus and soil pH (from Buckman and Brady, 1960)

nitrogen or potassium, phosphorus can, however, be very easily 'fixed' and hence rendered unavailable even in otherwise favourable soil conditions. The optimum soil pH range for availability is narrower than for the other macronutrients, and in even slightly acid or alkaline conditions it can form insoluble compounds with iron, aluminium and manganese (see Fig. 5.5). It can also be converted into an organic form in which it is temporarily unavailable. So, while removal by crops is low compared with nitrogen and potassium, high inputs (which may be three to four times greater than that actually removed) are needed to compensate for that fixed or immobilized in an unavailable form in the soil (Brady, 1974).

POTASSIUM (see Fig. 5.4)

Potassium fertilizers are also derived from mineral rock deposits. The most important forms are *potassium chloride* and *potassium sulphate* respectively (see Table 5.7). Although more abundant in most soils than any of the other nutrients, potassium resembles phosphorus in its relative unavailability. It is more akin to nitrogen in its susceptibility to leaching. Most of the 'native' potassium is held in unweathered rock minerals, while that added as fertilizer to the soil can be fixed in the lattices of certain clay minerals (particularly the montmorillionite and illite groups) from which it is only slowly released. Loss by leaching may equal that removed by crops, especially in tropical soils in which the adsorptive and nutrient-fixing capacity of the clay content is low. However, uptake by and concentration of potassium in the crop is high because of what is referred to as 'luxury consumption', i.e. uptake well in excess of crop needs.

CALCIUM AND MAGNESIUM

Other macronutrients are the so-called 'liming fertilizers' calcium and magnesium. These are derived from rocks such as limestone or chalk and are applied as: burnt or *quick-lime* (CaO combined with magnesium); hydrated or *slaked lime* (Ca(OH)$_2$); or ground rock (CaCO$_3$). The amount of calcium in the soil in general is large relative to demand. While it is an essential nutrient, it is also needed to maintain the soil at the optimum pH for nutrient availability and uptake (*c.* pH 6.5). Calcium cations are the most abundant of the neutralizing 'liming' bases (which also include magnesium and sulphur). They can counteract the acidifying effect of (SO$_4$)$^{2-}$, (NO$_3$)$^-$ and (CO$_3$)$^{2-}$ anions released into the soil solution by nitrogen fertilizers (e.g.(NO$_3$)$^-$) or by wet or dry deposition from the atmosphere (e.g.(NO$_3$)$^-$, (SO$_4$)$^{2-}$). The latter are particularly important in humid areas where, for part or all of the year, precipitation exceeds evapotranspiration. In the former case Ca^{2+} ions are easily displaced from the colloidal complex into the soil water by H$^+$ ions. In both cases the calcium is very easily leached out of the soil in drainage water.

Although there may be sufficient calcium for nutrition in the soil, increased acidity consequent on leaching often has serious consequences for the availability of other nutrients, or the mobility of non-essential toxic elements. The problem of combating acidity is most pronounced in many tropical soils under high-rainfall regimes. Much of the cation exchange capacity may be taken up by aluminium. This is precipitated by liming and becomes toxic at concentrations in excess of 1 ppm. Toxicity, however, is dependent on the species or variety of crop: many tropical crops are, in fact, adapted to acid soils; even within one species some varieties may be more or less acid-tolerant than others.

SULPHUR

It is only relatively recently that sulphur deficiencies in agricultural soils have been detected (Cooke, 1972). Although sulphur occurs in only small amounts in most soils, crop demand in comparison to those for the other macronutrients are not large; and since the mid-nineteenth century considerable quantities of sulphur have been added to soils in fertilizers such as superphoshate and ammonium sulphate, and from the industrially polluted atmosphere. However, modern cultivation of high-yielding crops together with the increasing use of fertilizers low in sulphur have resulted in deficiencies in some areas. More attention is now being given to the sulphur status of the soil and the application of specific sulphur fertilizers.

MICRO-NUTRIENTS

In comparison to the macronutrients the 'micros' are required in only very small quantities. Of the eight essential micronutrients, crop needs are highest for iron and chlorine, lowest for copper and molybdenum, and intermediate for manganese and zinc. Both the total amount and availability of particular elements is extremely variable. The former is a function of the chemical composition of the soil parent material and, in most instances, this relationship (established by modern soil surveying) is now sufficiently well known that deficiencies of micronutrients can be easily predicted. The latter is dependent on soil condition of which pH and the relative amounts of other macro- or micronutrients are the most important.

In the presence of a low pH, iron and manganese become more soluble and hence available (the latter to a degree that can be toxic), and molybdenum becomes less available. In more alkaline conditions iron and manganese tend to become 'fixed' in less-available forms. Those crops with high iron requirements may show deficiencies when the pH is as low as 5.0. Availability can also be affected by antagonistic reactions between particular micronutrients and other elements. For instance uptake of boron is curtailed in calcium-rich soils; zinc and iron when there is heavy use of phosphate fertilizer; and iron when copper is in excess or when manganese is in abnormally high or low concentrations in the soil. Rectifying deficiencies of micronutrients involves more diverse materials and methods of fertilization than do macronutrients. The former are frequently applied as dust or sprays (often mixed with pesticides) or combined with standard fertilizers or liming elements. However, deficiencies tend to be more serious for the livestock (and humans) than for the crop on which they feed.

NUTRIENT CYCLES IN AGRO-ECOSYSTEMS

In contrast to unmanaged ecosystems, agro-ecosystems are, in general, characterized by a larger and more rapid 'turnover' of nutrients. High outputs in crops and livestock are compensated by high inputs of fertilizers. The magnitude, rate and complexity of the cycle, however, depends on the type and intensity of the particular farming system. Frissel (1978) has categorized agro-ecosystems using nitrogen output of consumables as an index of *intensity* of farming. Nitrogen was chosen as the most important and characteristic element involved in the nutrient cycle. Ranking types of farming on this basis (see Fig. 5.6) illustrates an increase in

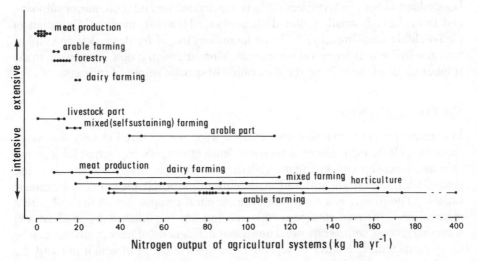

Fig. 5.6 Nitrogen output of agricultural systems. Black dots represent farm systems for which data is available (redrawn from Frissel, 1978)

nitrogen output with increasing intensity of inputs. The range of variation, however, is very much greater in the intensive than in the less-intensive systems. In the latter the level of management is low and the number and possible combination of inputs are negligible; in the former the level of management is highly sophisticated and the number and possible combinations of inputs are considerable. At one end of the spectrum the agro-ecosystem is little modified; at the other it is almost completely manipulated by man.

In contrast, however, the nutrient cycles in the least and most intensive agro-ecosystems are often less complex, because of a higher degree of specialization in input than in the 'intermediate' and characteristically mixed less-specialized systems (see Table 5.8).

Table 5.8 Comparison of annual gains and losses of nitrogen, phosphorus and potassium in the total soil compartment in ecosystems of varying intensity in North-West Europe

System	Nitrogen (kg ha^{-1})			Phosphorus (kg ha^{-1})			Potassium (kg ha^{-1})		
	Gain	Loss	Bal.	Gain	Loss	Bal.	Gain	Loss	Bal.
Traditional hill-sheep (UK)*	63	61	+2	4.3	4.6	−0.3	30	36	−6
Mixed self-sustaining farming (France)†	380	310	+70	81	39	+42	354	284	+70
Intensive winter wheat (UK)	138	123	+15	29.7	21.7	+8.0	128	163	−35
Intensive dairy‡	1076	724	+352	122	57	+65	515	391	+124

* Over 300 m altitude; semi-natural moorland; 0.5–1 sheep per hectare.
† Leys in rotation with arable crop; 75 per cent herbage used for grazing.
‡ Dutch dairy farm on clay soil; 4 cows per hectare; fertilizers and large amount of supplemental feed used; milk produced: 18 000 litre ha^{-1}; meat: 768 kg ha^{-1}.

(from Frissel, 1978)

EXTENSIVE AGRO-ECOSYSTEMS

Extensive agro-ecosystems are those in which the annual output of consumable nitrogen is less than 20 kg ha^{-1} (see Fig. 5.6). The output of crops and/or livestock per unit area is low and is dependent to a large extent on the natural or little-modified soil nutrient reservoir. They are represented by the traditional and relatively unsophisticated, but still widespread, systems of *livestock grazing* and *shifting agriculture* respectively. The former is associated with areas where physical constraints limit crop production; the latter with one of the earliest and most widespread methods of agriculture, still practised, mainly in the humid Tropics.

Extensive *livestock-grazing* systems are those which produce meat and other livestock products on the basis of the fodder provided by wild (uncultivated) vegetation (see Fig. 5.7). Fertilizers are not normally used. The only inputs of nutrients are those from the atmosphere and, of particular importance, those resulting from nitrogen fixation by free living or symbiotic bacteria in the soil. Loss of nitrogen by volatilization of ammonia from excrement and by denitrification, particularly in poorly drained soils, may be relatively high; that of potassium and phosphorus, in the meat and/or milk produced, is small. A most important and universal type of management used in these farming systems is burning, on a shorter or longer rotation, which aims to reduce the amount of dead, fibrous or woody vegetation and stimulate the production of fresh green tissues. Burning serves to speed up the nutrient turnover but it can also increase the loss of nitrogen and (in very high temperatures) of phosphorus by volatilization, and of all nutrients by the removal of ash by wind or water (see Table 5.9). However, work in Britain on heather-dominated moorland vegetation burned on a long rotation suggests that a balance of inputs to outputs can be maintained under a system of carefully controlled burning (Gimingham, 1975). A high proportion of nutrients are recycled via plant material within the system because a relatively small amount of the available fodder is consumed – the stock-carrying capacity being determined by minimum rather than maximum production levels. Also, a high proportion of the nutrients withdrawn are returned in livestock excrement. The nutrient cycle in this type of system is small and slow, and is highly dependent on the release of nutrients by organic decomposition, the rate of which is impeded by aridity, low temperatures, waterlogged conditions or low pH values in the soil. Nitrification is particularly slow and losses are therefore all the more serious.

In contrast, nutrient cycling in the system of *shifting agriculture* is characterized by a wide range of variation in volume and rate. The often very large natural biomass of tropical forest is cleared from a small area by felling, and the ground is prepared for cultivation by burning. The latter releases the nutrients which have accumulated in the wood over a long period of time (see Table 5.9). The resultant ash provides a nutrient-rich seed-bed in which a diverse and complete crop cover can be rapidly established and which helps to check loss by leaching and/or soil erosion. Loss of nutrients during burning and with the initiation of cultivation is relatively rapid. It has been estimated that approximately 20 per cent of the total nitrogen can be volatilized (Sanchez, 1976), but it is not known how much of this is returned in precipitation. Exchangeable bases and phosphorus decline and soil organic matter can decrease by 30–40 per cent in the first year of cultivation (see Fig. 5.8). After 2 or 3 years yields decline and the area is abandoned. With the re-establishment of the forest cover, nutrient storage in the biomass and the soil pool eventually restores the 'natural' balance.

The alternate exploitation and natural regeneration of the existing nutrient cycle was at the basis of many early developed settled agricultural systems in Europe. Cereal cropping to provide subsistence food was rotated with grazed fallow. There was comparatively little loss of nutrients from the system and the nutrient cycle was maintained, albeit at a low level. The manipulation of the 'natural' nutrient cycle by bush or grass fallows, which are grazed and not infrequently burned before cultivation, is still widespread, particularly in the subhumid to semi-arid areas of the Tropics. Here, however, pressure of population has resulted in a drastic shortening

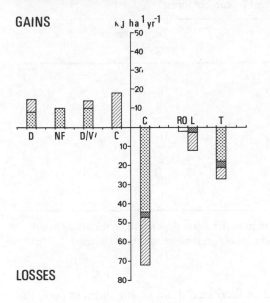

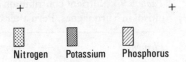

A NUTRIENT BUDGET

GAINS   kJ ha⁻¹ yr⁻¹

| | Nitrogen | Potassium | Phosphorus |

C   uptake by plant
D   droppings
dN   denitrification
D/W dry/wet deposition
F   fertilization
IM   immobilization in organic matter
L   leaching
M   mineralization
MF   mineral fixation
MO   mineralization in organic matter
NF   nitrogen fixation
RO   run off
V   volatilization
T   total

LOSSES

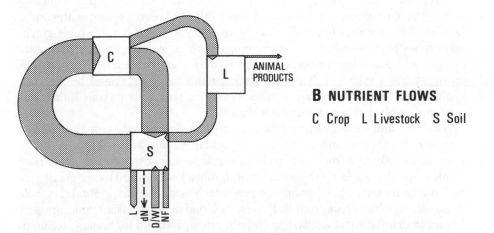

B NUTRIENT FLOWS

C Crop   L Livestock   S Soil

Fig. 5.7 Nutrient budget and nutrient flows of available nutrients in a traditional hill farming system in the UK: flows proportional to amounts of nutrients moving between soil, crop and livestock pools. Diagrams constructed from data (estimates) given in Frissell (1978) (from Tivy, 1987)

Table 5.9 Nutrient contribution of ash and partially burned material deposited on an ultisol in Yurimaguas, Peru after burning a 17-year-old forest*

Element	Composition (%)	Total additions ($kg\ ha^{-1}$)
N	1.72	67
P	0.14	6
K	0.97	38
Ca	1.92	75
Mg	0.41	16
Fe	0.19	7.6
Mn	0.19	7.3
Zn	132 p.p.m.	0.3
Cu	79 p.p.m.	0.3

* It should be noted that the nutrient content of the ash is dependent on variations in soil properties, clearing techniques and percentage of forest biomass actually burned.

(from Sanchez and Salinas, 1981)

of the regenerative fallow period and an increased drain on the nutrient pool, the consequences of which will be discussed in a later chapter.

INTENSIVE AGRO-ECOSYSTEMS

Intensive highly specialized *arable* or *livestock* agro-ecosystems are those in which very high outputs per unit area are maintained by a commensurably large input of nutrients. Both the volume and the rate of cycling is high. In the continuous and wholly *arable* systems (see Fig. 5.9) without ley grass, nutrient input is predominantly in the form of inorganic fertilizers (see Table 5.10). Sources of organic matter on the farm are limited mainly to cereal, roots, straw or stubble, and in some cases green-manure supplied by wastes of arable crops such as sugar-beet tops. Farmyard manure and other organic fertilizers are seldom used because of high costs of transport and application. Nitrogen fixation tends to be depressed by the large applications of nitrogen fertilizers, while organic matter, except on soils initially rich in humus, is often at its minimum stable level.

The exact amount of nutrients taken up varies according to the crop varieties and combinations thereof and to the type of system within which they are grown. However, uptake of nitrogen and potassium relative to phosphorus is high in most arable crops (see Table 5.11) – particularly in roots. Potatoes and forage grasses are high nitrogen users, while leguminous crops such as soya beans, groundnuts, field beans etc., capable of fixing nitrogen, do not respond to inputs of inorganic nitrogen. Losses from intensive arable systems, where crops produced for human consumption or fodder to be sold off the farm, can be very great. That not 'exported' in the crop – the *'residual nutrients'* in the soil, including much of the nitrogen and nearly all the available potassium and phosphorus – can be lost by leaching during that part of the year when the soil is bare.

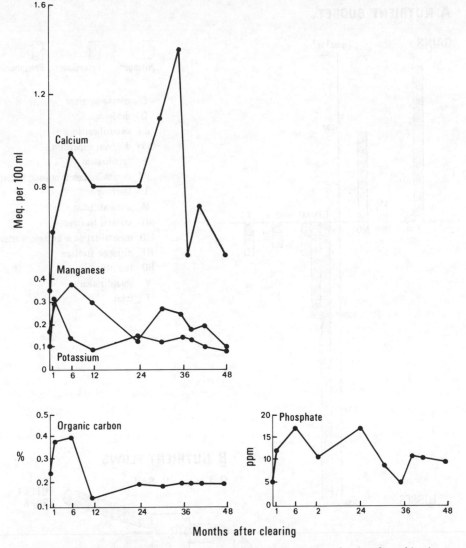

Fig. 5.8 Changes in nutrient content of tropical forest soils after clearing for cultivation (from Sanchez and Salinas, 1981)

The nutrient cycle in the *intensive livestock* system (see Fig. 5.10) differs from that in arable systems in several important respects. Particularly in those based on grass production, with little or no arable acreage, inputs of nitrogen can be exceptionally high (see Tables 5.10 and 5.11) in order to ensure two or more cuts for silage during the growing season, in addition to grazing. Inputs from organic sources such as FYM and, increasingly, slurries, tend to be very important. Uptake of nutrients by livestock via the fodder can also be high, particularly in the case of young animals and dairy cows. Much, however, is returned in their excrement, which contains most of the plant foods in the hay and silage which has been fed indoors (Cooke, 1972),

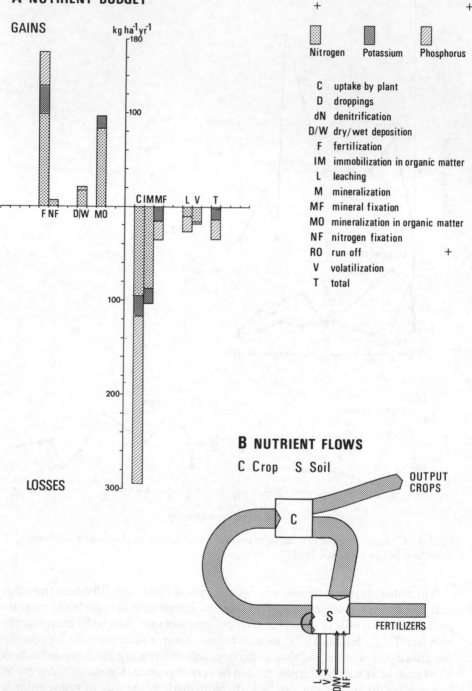

Fig. 5.9 Nutrient budget and nutrient flows of available nutrients in an arable monoculture, winter wheat in the UK (see fig. 5.7 for explanation) (from Tivy, 1987)

Table 5.10 Nitrogen economy of arable crops and temporary grassland (leys) in the UK, 1974

	Input	Nitrogen (kg tonne)	Output	Nitrogen (kg tonne)
Arable crops	Fertilizers	370	In crops	460
	Farmyard manure	30	Leaching	120
	Total	400	*Total*	580
	(extra from biological sources)	370		
Temporary grassland	Fertilizers	220	In crops	440
	Natural sources	450	Leaching	10
	Total	670	*Total*	450
	Farmyard manure	70		
	Excreta	200		
	Total	270		

(from Cooke, 1977)

Table 5.11 Nutrients in crops grown, and in excrement of livestock on UK farms, 1973

	Nutrients (kg tonne)		
	N	P	K
Crops			
Arable crops	460	75	500
Grass	1100	170	1000
Excrement			
Cattle	600	150	540
Sheep	150	30	120
Pigs	30	6	10
Poultry	60	12	20
Estimate of total amounts dropped on grassland	500	120	450

(from Cooke, 1977)

and is often further enriched by the supplementary concentrates necessary to maintain high milk production. In this case the manure may contain suffient nutrients to obviate the need for inorganic fertilizers. Loss of nutrients by leaching, unless supplied in amounts and at rates greater than can be absorbed by the plants, is curtailed by an all-year grass cover, which in temperate climates may be capable of some growth even in winter. That by volatilization and/or denitrification of nitrogen

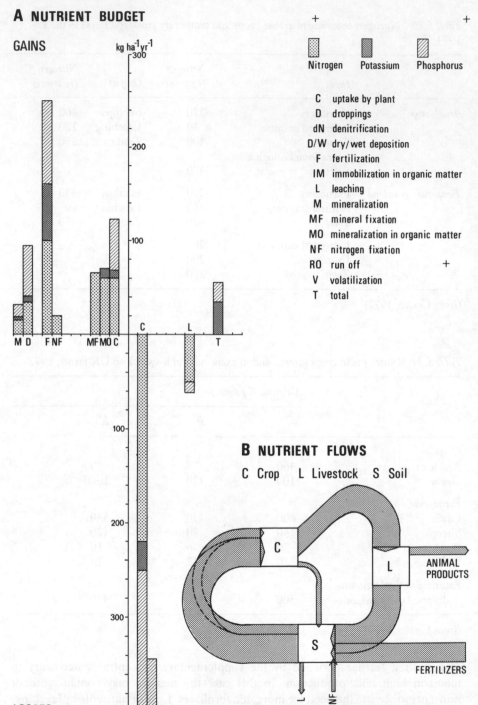

A NUTRIENT BUDGET

GAINS   kg ha⁻¹yr⁻¹

☐ Nitrogen   ☐ Potassium   ☐ Phosphorus

C   uptake by plant
D   droppings
dN   denitrification
D/W dry/wet deposition
F   fertilization
IM   immobilization in organic matter
L   leaching
M   mineralization
MF   mineral fixation
MO   mineralization in organic matter
NF   nitrogen fixation
RO   run off
V   volatilization
T   total

M D   F NF   MF MO C   C   L   T

LOSSES

B NUTRIENT FLOWS

C Crop   L Livestock   S Soil

Fig. 5.10 Nutrient budget and nutrient flows of available nutrients in a mixed farming system (see fig. 5.7 for explanation) (from Tivy, 1987)

is greater than in arable systems and, depending on the type of grass and particularly its rooting system, higher levels of organic matter are maintained than under arable cropping. Although cereal crops may produce as great a volume of roots as grass in one year, the former decompose more rapidly than the latter (Russell, 1977).

MODERATELY INTENSIVE AGRO-ECOSYSTEM

The traditional and formerly more extensive *mixed farms* (see Fig. 5.11), character-ized by livestock production on the basis of arable fodder and food crops and improved grassland, are intermediate between the extensive and intensive systems defined by Frissell (1978). Nutrient inputs and cycling are dependent on animal excrement, FYM and other wastes produced on the farm; and on the soil organic-matter reserves. In the former, a high proportion of the nutrients – most of the phosphorus, calcium, magnesium and nearly all the potassium – in the grass consumed by livestock is returned in the excreta. On continuously grazed grassland this is dropped unevenly in large 'pats' from which losses of ammonia by volatiliza-tion, and of other minerals as a result of surface run-off or wind erosion, can be high. Farmyard manure, evenly spread prior to the growing season, ensures more effec-tive nutrient conservation; and those traditional systems where animals are kept indoors at night or all day in the winter, are bedded on straw and are fed on hay or silage and other arable fodder crops grown on the farm provide the ideal conditions for the inexpensive production and use of FYM as a source of nutrients.

The maintenance of a soil organic-matter nutrient store is dependent on the use of ley (rotation) clover/lucerne pastures, in which nitrogen fixation is an important input. For this reason the leys form what have been referred to as the 'nitrogen-catchment' part of this particular type of agro-ecosystem. They have also been used in the arable rotation as a 'break crop', helping to control soil-borne diseases, to build up organic matter and to rehabilitate the soil structure, which may have deteriorated with crop cultivations. The ratio of arable to grass is dependent on the type of livestock system and the duration of the ley pastures on the one hand, and on the extent to which environmental conditions are more or less suitable for crop production on the other.

The development of the ley-farming system in Britain in the late-eighteenth and nineteenth centuries was basic to the improvement of agriculture in terms of not only increasing but maintaining output by skilful management of the nutrient cycle. It was primarily dependent on the increased inputs from organic sources, though it also benefited from the supplies of inorganic fertilizers that became available at this time, such as basic slag and superphosphates. It was common in Britain in the nineteenth century to apply NPK fertilizer to the extremely demanding root crops (swedes/turnips) every fourth year, with the following cereal crop making use of the *residual nutrients* left in the soil. The control of the nutrient cycle is also dependent on the rotation not only of grass and arable crops but of arable crops with varying nutrient demands so that the losses from one would be counteracted by the gains from another.

In general the 'characteristic' traditional mixed system is one of moderate to low inputs of nutrients from outside the farm. Losses by denitrification, leaching and

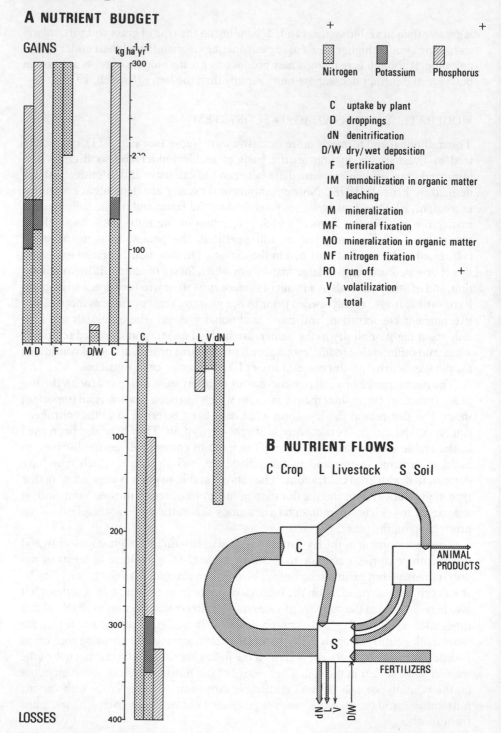

A NUTRIENT BUDGET

GAINS

kg ha⁻¹yr⁻¹ $kg\ ha^{-1}yr^{-1}$

Nitrogen   Potassium   Phosphorus

C   uptake by plant
D   droppings
dN denitrification
D/W dry/wet deposition
F   fertilization
IM immobilization in organic matter
L   leaching
M   mineralization
MF mineral fixation
MO mineralization in organic matter
NF nitrogen fixation
RO run off
V   volatilization
T   total

M D   F   D/W C   C   L V dN

B NUTRIENT FLOWS

C Crop   L Livestock   S Soil

LOSSES

Fig. 5.11 Nutrient budget and nutrient flows of available nutrients in a dairy system, Netherlands (see fig. 5.7 for explanation) (from Tivy, 1987)

volatilization are also small, particularly on farms where only livestock are 'exported'. The volume and rate of the cycle are 'intermediate' between those of the extensive and intensive systems. However, in 'model' form it closely approximates to an almost closed, local or farm-based, self-sustaining nutrient-conserving cycle.

In reality the range of variation within each of the three categories of agricultural nutrient cycle is dependent on whether the data are estimated or calculated; and on the varying combinations of nutrient inputs and outputs which are a function of both type of management and environmental conditions. As with all ecosystems, agro-ecosystems can never be completely closed. In the latter the output from one frequently becomes the input of another. Further, in terms of nutrient cycling, agro-ecosystems vary in scale from the individual farm (or even sub-farm) or working unit to that which involves the whole of the ecosphere and in which, because of extensive trade in agricultural products, the nutrient routes are of global dimensions.

Not only does the type of nutrient cycle vary but so does the nutrient balance. While the volume of the soil cycle has been increasing in those areas of the world which are intensively farmed, an equilibrium has been achieved between gains and losses for most nutrients with the exception, in some areas, of phosphorus. Cooke (1972) remarks that an outstanding feature of UK soils is that they are gaining phosphorus and have been doing so for many years. In many parts of the world, particularly where extensive arable and largely subsistence farming is practised, nutrient inputs are low (because of the shortage of organic wastes and the expense of inorganic fertilizers) and nutrient volume is small and net losses are increasing. The paucity of nutrient inputs into the agro-ecosystems of the poor developing areas contrasts markedly with the abundance of the richer intensively farmed. Until relatively recently achievement of high agricultural productivity has been at the cost of high and often, it is now realized, wasteful inputs. Cooke (1977) maintains that *direct waste* of fertilizers occurs when more is applied than crops need; and when the correct amount is applied at the wrong time or place and in forms unsuitable for the crops and/or soil. *Indirect waste* occurs when natural losses are increased by farming methods and there is a failure to make full use of soil reserves as well as of crops in which nitrogen is biologically fixed. Also, many do not allow for residual nutrients in arable systems or for differences in the way grassland is managed. Waste results in a lower ratio of nutrient gains to losses, and in inefficiency of fertiliser use, which will eventually reflect on the agricultural productivity of the system.

Chapter 6

Agricultural productivity

In both unmanaged ecosystems and agro-ecosystems biological productivity is expressed in terms of the rate of plant and/or animal biomass accumulated per unit land area within a specified time period. In both it is a function of the same basic process – photosynthesis – whereby simple inorganic elements (carbon, oxygen, hydrogen, nitrogen, potassium, phosphorus) derived from the atmosphere and the soil are converted, by chlorophyll-carrying plant cells using light energy, into complex organic compounds (carbohydrates, proteins, fats). In both types of ecosystem the rate of plant growth (*net primary productivity*; NPP) is dependent, on the one hand, on the efficiency with which the available solar radiation is intercepted and used; and, on the other hand, on the difference between the rate of photosynthesis (*gross primary productivity*; GPP) and the rate of *respiration* (R) during which the energy used in plant metabolism is dissipated as heat, i.e.

$$NPP = GPP - R$$

Net primary production is usually recorded as the weight of dry matter production per unit area per unit time.

Net primary production in any ecosystem provides the food base for secondary (consumers and decomposers) production. One of the principal aims of agriculture is to channel as much as possible of the energy from incoming solar radiation into selected crops and/or livestock, and to minimize that used by such potential competitors as weeds and pests. As a result food-chains are shorter and the resulting food-web is simpler in most agro-ecosystems than in completely unmanaged systems. It is, however, very difficult and not very illuminating, to compare the net primary productivity of the unmanaged with that of the agricultural ecosystem. First, there are few measures of total biological productivity, including root biomass, for crops. Secondly, productivity or *yield* in agro-ecosystems refers to the *utilizable part* of the plant, which is not the same in every crop. Further, the utilizable part is always less than the total biological production (see Table 6.1). Then, as Loomis and Gerakis (1975) point out, not only are agricultural yields very variable, they rarely reflect exactly either the crop or the environmental potential. This is because of the cultural and economic constraints on crop choice and the need to minimize risks.

HARVEST OR CROP INDEX

The proportion of the 'recoverable' crop yield which contributes to the final *utilizable* or *commercial yield* is commonly known as the *harvest* or *crop index*, i.e. the ratio of commercial to recoverable yield. In the case of cereals this may be expressed as the grain to straw or the grain to stover (maize) ratio. The harvest index varies widely depending on whether crop yield is a function of a vegetative or a repro- ductive stage of growth (Table 6.2). In the former case, which applies to tubers, roots, green vegetables and forage grasses, the harvest index is usually relatively high. It can range from 85 per cent in main-crop potatoes, close to recoverable yield in forage grasses, to less than 50 per cent in cereals, grain-legumes, cotton and oil- seeds (15–25 per cent). The index can also vary within a particular cultivar or crop strain depending on density of planting, or on variations in the supply of nutrients and water. In addition the index is usually based on the *harvest at maturity*, which is normally less than that immediately before harvest, because of respiration and leaf losses. Finally, the harvest index can be a function of management. Indeed, there has been a marked increase in recent decades in the index of cereal cultivars, without a significant increase in their total recoverable yield (see Table 6.3). This is a result of a number of factors. One is the use of growth-regulating hormones, such as the chemical chlormequat, to produce a shorter stem. Another is deliberate breeding of crops with shorter stems, reduced tillering and larger flower-heads which mature earlier. The extent to which the stem and/or leaf can be reduced is limited by the minimum requirements of the plant for mechanical support and by the optimum leaf-area index for light interception. Reduced plant height also makes for greater susceptibility to disease, poor threshability and lower competitiveness with weeds (Sharma and Smith, 1986).

CROP YIELD

Crop yield is a function of a more complex set of interacting variables than is primary biological production in the unmanaged ecosystem. These include:

1. The environmental conditions under which the crop is grown.
2. The yield potential of the particular cultivar.
3. The management of the crop and its environment in order to minimize environ- mental constraints on the realization of the maximum yield potential of the crop.

   The *environmental conditions* that control the rate of growth and the accumulation of organic matter are the same in all plants (see Fig. 6.1). Of these, the amount of incoming solar radiation available for photosynthesis and the efficiency with which it can be intercepted and used are usually considered to be the ultimate factors determining the maximum primary biological productivity that can be achieved.

Table 6.1 Contribution of various vegetative organs (minus roots) of three cereals to biological yield

	Percentage biological yield		
	Winter wheat	*Spring barley*	*Oats*
Leaves	9	6	7
Stems	33	28	34
Vegetative tillers	8	5	3
Chaff	10	10	15
Grain	40	51	41

(from Donald and Hamblin, 1976)

Table 6.2 Approximate proportion of crop that is represented in harvestable yield

	Harvestable yield (%)*
Wheat	54–60
Perennial ryegrass (at one harvest)	63
Maize	42
Field peas	50
Lettuce	50–90
Brussel sprouts	30

* Harvestable yield dry matter is presented as a percentage of total above-ground plant dry matter.

(from Spedding, 1975)

Table 6.3 Harvest index in old and new cereal cultivars*

	Crop components of yield		
Crop and location	*YR (tonne ha^{-1})*	*YC (tonne ha^{-1})*	*Harvest index (%)*
Wheat (UK)	11.00 (12.72)	2.59 (4.38)	23.50 (34.40)
	8.56 (10.02)	2.35 (3.61)	29.69 (36.00)
Barley (UK)	11.04 (10.86)	4.38 (5.21)	39.71 (48.00)
Rice (Philippines)	16.50 (17.50)	2.76 (4.18)	16.80 (24.10)
Maize (Uganda)	17.70 (18.66)	2.49 (3.69)	14.00 (19.70)
Sorghum (N. Nigeria)	39.90 (13.70)	2.66 (4.81)	7.10 (35.10)

* Results for new varieties given in parentheses.
YR = recoverable biological yield; YC = grain yield.

(from Holliday, 1976)

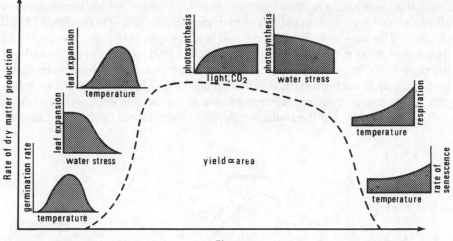

Fig. 6.1 Hypothetical relationship between the seasonal rate of dry-matter production of a crop, the physiological processes involved and environmental conditions (redrawn from Monteith, 1965)

However, temperate plants grown in full sunlight generally attain their maximum photosynthetic rate before light saturation is reached. This is because the supply of atmospheric carbon dioxide is not sufficient to allow maximum light use. It has been suggested (Monteith, 1972b) that the recent increase in atmospheric carbon dioxide

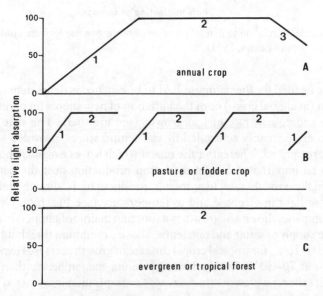

Fig. 6.2 Schematic representation of growth stages in different kinds of crops: 1) Stage in which part of incoming light energy reaches the soil; 2) Stage of closed green crop; 3) Stage of senescence (leaf discoloration or leaf fall) (from Alberda, 1962)

concentration resulting from the combustion of coal, oil and gas has been such as to allow a potential increase in the rate of biological production of the order of 15–20 per cent. The amount of light intercepted is a function of the size, structure and duration of the crop canopy (Monteith and Elston, 1971) and, more particularly, of the ratio of total leaf surface area to ground surface area, i.e. the *leaf-area index* (LAI) (see Fig. 6.2). It has been calculated that indices of 4–7 (according to the morphology of the crop) are required to intercept most of the incident light; and, at a LAI of 4–5, over 80 per cent of the available light will be intercepted by the crop canopy.

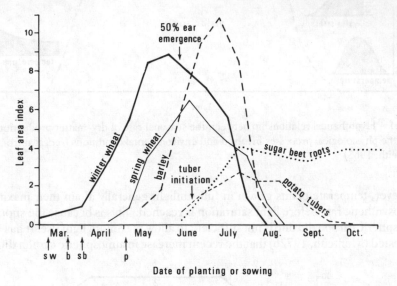

Fig. 6.3 Changes in relation to date of planting or sowing in the leaf-area index (LAI) of different crops (from Watson, 1971)

The time needed for the optimum LAI to be reached is dependent on the rate of germination (in annual crops) or of the initiation of new shoots (in perennial crops), and on the subsequent rate of leaf growth (see Fig. 6.3). The date and rate of germination are primarily controlled by soil temperatures, provided soil moisture conditions are optimal. Thereafter the rate at which leaves expand, when the LAI is low, is also an important factor in total crop production over the whole growing season. Initially growth rate (dry-matter production) is directly related to the percentage of light intercepted and to temperature (see Fig. 6.4). Once the crop canopy is complete, the rate of growth is a function mainly of temperature, assuming an adequate supply of water and nutrients. Above a minimum threshold (0.05 °C for temperate, 10–15 °C for tropical crops) the rate of growth increases exponentially to an optimum at 20–30 °C (with a lower optimum for temperate than for tropical crops). Thereafter the growth rate decreases as respiration increases at a rate which exceeds that of photosynthesis.

Total biological production by the crop is dependent on the extent to which a closed canopy can be maintained during the climatically favourable growing season.

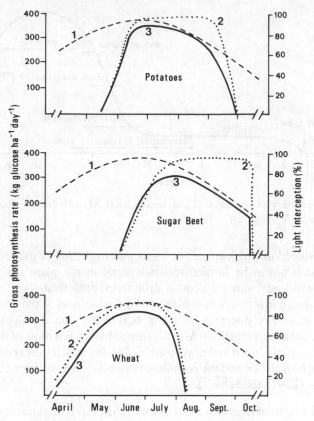

Fig. 6.4 1) Potential gross photosynthetic-rate curve; 2) Percentage light-interception curve; 3) calculated actual gross photosynthetic rate curve for potatoes, sugar-beet and wheat (from Sibma, 1977)

In many crops, and particularly those with a determinate growth habit, leaf production stops just before flowering (see Fig. 6.3) and thereafter photosynthesis depends on the persistence (duration) of existing green leaves. In contrast, indeterminate plants such as potatoes and sugar beet can continue producing new leaves for as long as the growing-season conditions are favourable. The final utilizable or economic yield then depends on how much of the assimilated material is produced in, and/or accumulated by, the yield organ.

On the basis of where and when the economic yield is finally located, Bunting (1975) distinguished three phenological categories of crop plants:

1. Yield produced throughout much or all of the growing season because it consists of the vegetative parts of a perennial or biennial crop (e.g. fodder grasses and other forage or silage crops, sugar cane, many roots and tuberous crops, taro, cocoa, rubber, tea, sago). In these essentially 'vegetative' crops the accumulation of yield in the source organs (i.e. leaves, stem) or in storage organs (e.g. tubers, roots etc.) can proceed over a long, indeed an almost indefinite, period provided growing conditions are suitable.

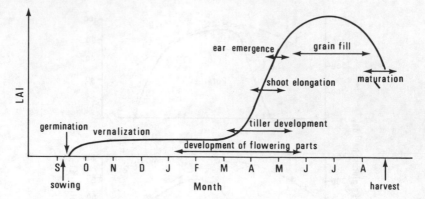

Fig. 6.5 Schematic representation of leaf-area index (LAI) with growth phases in a cereal crop (supplied by A. Williams)

2. Yield produced during a greater or lesser part of the life of the crop in fruits or seeds that begin to be formed relatively early in the plant's life (e.g. grain-legumes, oil-seeds, tomatoes, cotton, fruit trees, bush fruits).
3. Yield produced in terminal or late-formed inflorescences as the final phase in the life of an annual crop (cereals) (see Fig. 6.5) or the annual shoot in a perennial crop (e.g. bananas, plantains). As is illustrated in Fig. 6.6 most of the yield in the small grains is produced in the last-formed *flag-leaf* and in the ear over a period of only some 6–8 weeks; and the economic yield is half or less (6 tonne ha^{-1}) that in root crops (12–15 tonne ha^{-1}).

The total weight of dry matter laid down in the crop is dependent on the size of the photosynthesizing surface (LAI), while the rate of production of the economically important yield organs (e.g. the harvest index) is a function of the length of the yield-forming phase. To achieve maximum yield in a given thermal regime, the crop must have an adequate supply of water and plant nutrients during its period of growth and development. A deficiency of either can limit potential yields.

LIMITING FACTORS

Water is considered the single most important factor limiting crop yields on a global scale; and agriculture is still the major 'consumer' of water in the world today. Maximum photosynthesis occurs when the plant stomata are wide open, a condition dependent on a continuous supply of water to keep the guard cells turgid, and is normally attained when soil water is near, but just below, field capacity, i.e. when the soil water deficit is at a small but finite value of approximately 2.5 cm (Monteith, 1977). Above this threshold, leaching of nutrients or poor soil aeration can become limiting. A water deficiency which checks early growth and canopy development can curtail the total biological production over the whole growing season. Potatoes require the soil to be at field capacity during the period of development of underground organs; in the determinate cereals, whose yield is dependent on the

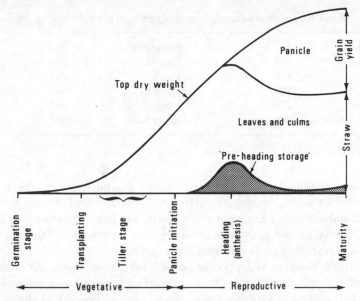

Fig. 6.6 Growth stages in rice: changes in the amount of temporarily stored carbohydrates (preheading storage) and the dry weight of various parts according to growth stages (from Murata and Matsushima, 1975)

number of grains per unit area, the critical growth phase when yield can be seriously checked by a water deficit is just before and during flower formation (i.e. anthesis). For those crops whose economic value is a function more of fresh than of dry weight, a water deficit during their production phase may be sufficient to effect a reduction in the size (or weight) class by which they may be graded for sale.

The efficiency with which crops use water is usually expressed as the ratio of dry weight produced to water used (see Table 6.4). While *water-use efficiency* varies with the other environmental conditions (including diseases) that affect yield, some crop varieties are more efficient in this respect than others, C4 crops generally being twice as efficient as C3 crops. One of the aims of agricultural management is to increase the water-use efficiency of field crops. Over the last 50 years efficiency has doubled in wheat, barley, rice and cotton and has increased five times in maize and soy. This is mainly the result of fertilizer application at a rate greater than the water consumption of the particular crop. Breeding early-maturing varieties or those adapted to cooler or drier growing conditions has also contributed to higher water-use efficiency; for example cotton that matures in 120–130 days after planting is 25 per cent more efficient than a variety that requires 150–180 days (Begg and Turner, 1976)

A deficiency of one or more nutrients can, under otherwise favourable conditions, limit yield; and the limiting nutrient will be that with the largest suboptimal deficiency. This is the basis of the *Law of the Minimum* first propounded by the agricultural chemist Leibig in the nineteenth century. There are still many parts of the developing world where low crop yields are related, in part, to deficiencies of phosphorus and nitrogen, particularly in old, highly weathered and leached soils

Table 6.4   Water requirement of crop plants (g water used per g dry matter produced)

Sorghum	322	Soya beans	744
Maize	368	Sweet clover	770
Wheat	513	Vetch	794
Oats	597	Potatoes	636
Rice	710	Cotton	646
Flax	905	Sugar beet	397

(from Spedding, 1975)

which have developed on acid igneous rocks, or sandstones. In the developed countries of the world, optimum nutrient levels on crop land are maintained by a continual and still increasing input of fertilizers, and more particularly of nitrogen; in many, potassium and phosphate levels are now roughly stable and the status of most soils in respect of these nutrients is adequate.

Soil fertility, however, is dependent not only on nutrient status but on the ability of the soil to supply nutrients and water to crop roots at a rate commensurate with uptake. The amount of available nutrients and water is dependent on the physical condition of the soil, particularly its cation exchange capacity and water-holding capacity. Management of the soil by tillage, fertilization, and drainage or irrigation aims to enhance, as far as possible, the physical fertility of the soil. Actual crop yields will depend on the degree of adaptation that can be achieved between the crop requirements for maximum yield and the environmental resources available.

There is also a close relationship between water stress and nutrient deficiency. Nutrient levels in the soil are usually highest near the surface, which is also that part of the soil which dries out first. Hence reduced growth when there is a moderate water deficiency may, in part, result from a reduction in uptake of nitrogen and phosphorus. Conversely, soil nutrient status can affect water use. High nitrogen levels stimulate vegetative growth – the demands of which may lead to soil water deficits that reduce yields to below those that might have occurred as a result of low nitrogen levels (Cooper, 1975).

Crop yield is, however, very variable both spatially and temporally and the two main causes of variation are:

1. The so-called 'negative' biological factors.
2. The weather.

The *negative biological factors* are the weeds, pests and pathogens – which comprise the third major organic component of all agricultural ecosystems. They are the unwanted and agriculturally unproductive 'enemies of the farmer'. More precisely, they are, from the farmer's point of view, plants (weeds), animals (pests) and pathogens which have the potential, when their populations reach or exceed a certain threshold, to reduce the quantity and/or quality of crop or animal yields and to increase costs of production. Weeds and pests depress yields either indirectly by competing for resources, reducing crop and livestock health, or directly by the consumption of agricultural products.

WEEDS

The number of weed species is very small when compared with the total world flora and a high proportion belong to a few families of flowering plants. In eastern North America seven families contribute 60 per cent of the 700 species of introduced weeds. While many are herbaceous, some of the rangeland weeds are woody.

Weeds of agricultural land have several sources of origin, including:

1. Open, disturbed habitats in which only those plants adapted to unstable conditions can survive, such as natural coastal or riverine habitats and man-made sites such as road and rail cuttings, ditch and canal banks; gravel workings etc.
2. Relatively undisturbed woodland or grassland habitats which contain species capable of growing more vigorously in agricultural habitats because of lack of competition for light.
3. Exotic species introduced accidentally or deliberately from other environments.
4. Cultivated species which have 'escaped' from gardens.
5. Crop-seed contaminated by weed seed.

The most successful weeds, which cause the greatest crop losses, are those plant species particularly well adapted to take advantage of and grow more vigorously in the man-made, man-disturbed agricultural habitats than they would elsewhere. While agricultural weed plants share many morphological and physiological characteristics, common to all is the ability to survive on open disturbed land. This facility is closely related to the high reproductive capacity combined with rapid growth and establishment, characteristic of the majority of weeds. Many are annuals with a short life-cycle, and a rapid and relatively high seed production (even under poor environmental conditions) which is spread over a long period. Dormancy is variable but, nevertheless, many weed seeds can remain viable for a relatively long time buried in a cultivated soil. Hill (1977) notes that 47 per cent of the seeds of shepherd's purse (*Capsella bursa-pastoris*) germinated after 16 years; 84 per cent of the greater plantain (*Plantago major*) after 21 years and 83 per cent of the black nightshade (*Solanum nigrum*) after 39 years. In addition, many weed seeds can survive and germinate successfully in open exposed sites subject to wide ranges of day and night temperatures.

Flexibility of seed production undoubtedly contributes to a high survival rate, even in the face of intraspecific competition, and to resistance to the rapid environmental changes common to the agricultural habitat. Many of the more serious and difficult weeds are perennial plants with particularly well-developed powers of vegetative reproduction from surface or underground food-storage organs such as stolons or rhizomes whose growth can be stimulated by fragmentation during cultivation. Some of the most successful perennial weeds are grasses, e.g. common couch or quake grass (*Agropyron repens*), Johnson grass (*Sorghum halpense*) and wild oat (*Avena fatua*).

Another characteristic of the successful weed is its adaptation for both long and short-distance seed and/or vegetative dispersal. This includes forms which facilitate transport by wind, water or attached to animals or within the faeces of animals. However, the close association of the weed and the crop plant has inevitably resulted in man being, directly or indirectly, the most important agent of weed dispersal today. Weeds are widely and rapidly dispersed as contaminants in crop seed, in hay, silage and other animal feedstuff, and in straw; on agricultural machinery and vehicles; in packaging materials; and in the bulk movement of sand, ballast and soils. A very considerable number of agricultural weeds (e.g. chickweed (*Stellaria media*), knotgrass (*Polygonum aviculare*), charlock (*Sinapsis arvensis*) and annual meadow grass (*Poa annua*)) now have a virtually cosmopolitan range. The efficiency of the weed dispersal, however, is largely a function of the viability and longevity of its seed.

Weeds in general exhibit a wide range of tolerance to variations in the physical environment and are well adapted to survive in disturbed cultivated habitats, fully exposed to extremes of temperature and to rapid variations in surface moisture conditions. Others can take advantage of the more favourable microclimate provided by the growing crop. Indeed the success of many weeds is related to the adaptation of their growth forms and development patterns to that of the crop they infest and, more especially, to their ability to survive the particular method of cultivation being used. Many show a high degree of 'mimicry' of the associated crop, e.g. grass weeds in cereals and pastures. An annual such as winter wild oat (*Avena ludiviciana*) can infest winter-sown wheat because it can become established before the wheat crop is tall enough to shade it out in the spring. Again, weeds with a short life-cycle can grow and produce seed before or even between cleaning and cultivation. In contrast, weeds of improved and/or cultivated hay pastureland are often avoided by livestock because, for one reason or another, they are unpalatable; and as a result they have a competitive advantage over the grazed plants. However, as Hill (1977) points out, the relative aggressiveness of a particular weed depends on the type, stage and management of the crop on the one hand and on the prevailing environmental conditions such as weather and soil type on the other.

It has been suggested that weeds may have beneficial aspects in their contribution to soil organic matter, reduction of soil erosion and concentration of nutrients which deeper rooted species may help to recycle. These benefits, however, would probably be more than outweighed, from an agronomic point of view, by the fact that they frequently harbour actual or potential crop pests. In fact most insect pests also feed on wild plants, particularly when the latter are closely related to the crop; and weeds provide a source of pest food particularly when the crop growing period is shorter than that of the insect feeding season and as a result ensure a reservoir of crop pests and disease.

PESTS AND PATHOGENS

The pests and pathogens of crop plants are the unwanted 'consumers' in all agro-ecosystems. The largest group of pests is that comprising the *insects* and *mites* of which the majority are plant-eaters. It has been estimated that 500–600 species attain pest status in the USA. Population numbers (particularly of insects) living in the soil and in the air are enormous – 25 million per hectare of soil and 22 million fly larvae per hectare of oats are not unusual figures. Reproduction rates and population growth can, given favourable conditions, be extremely rapid. The cabbage aphid, for example, has the potential to produce a new generation every 2 weeks.

Some of this group of pests are *generalists* in their feeding habits, others are more *specific*, attacking only a limited number of closely related plants. Indeed many of the pest insects common on cultivated plants in many parts of the world are monophagous. Some inflict direct damage or destruction, others are more significant as disease vectors. The economic importance of pest damage is closely related to the type of crop concerned. As Wigglesworth (1965) has pointed out, plants with an indeterminate flowering habit are generally subject to attack over a longer period than are those with a determinate habit. Hence crops like cotton, coffee and deciduous fruits require a greater amount of pest control than those such as maize or the small grains.

Another large group of pests is composed of the *nematodes*, microscopic non-segmented eelworms, which inhabit the soil and usually eat underground plant parts. They tend to be more prevalent in warmer climates, though the most important is probably that causing 'root-knot', which is cosmopolitan in range. Other pests are the *snails* and *slugs* and some *vertebrate* animals – mostly small mammals such as rodents and birds, though the emu and the red kangaroo, which have attained pest status in Australia, are among the larger animals in this category.

The *pathogens* are the disease-causing micro-organisms, which include:

1. *Fungi:* responsible for the greatest number and diversity of plant diseases. Most agricultural crops are susceptible to fungal disease and some can be infected by as many as thirty different species.
2. *Bacteria:* account for relatively few crop diseases though some have, at various times, caused exceptionally severe damage.
3. *Viruses:* infect almost all the higher plants.

The pests and pathogens of crop plants originate in much the same way as do weeds, either by co-evolution with the ancestors of modern crops or by deliberate or accidental transport from a known to an alien environment where the natural constraints on population numbers do not operate. In both instances, however, increase of a population to the level of an agricultural pest is related to a greatly increased supply of high-quality crop food grown under optimal climate conditions for pest reproduction, combined with a drastic reduction, or even absence, of natural predators. Reduction of natural enemies of the pest is usually the result of one or more of the following factors:

1. Introduction of the pest into an alien habitat where natural predators either do not exist or, if introduced, cannot tolerate the new environment.
2. Alternation of different crops (sometimes with fallow periods), which may create a time-lag between the introduction of a new crop and immigration of predators.
3. The greater susceptibility of predators to pesticides and other toxic substances used on the farm than that of the 'target' pests.
4. Natural predators may have alternate prey or adult food requirements other than crop pests.

In addition, high crop densities favour the even spread of comparably high pest populations before intraspecific competition and predator immigration can begin to take effect. A unique characteristic of crops is that they exhibit an almost universal specificity of particular pathogens; relatively few diseases affect different crop species. The crop pest/pathogen problem has increased with the modern trend in arable farming towards long-term monocultures, particularly of crops with a relatively long maturation period, a high degree of genetic uniformity and where climate and/or irrigation favour continuous cultivation of the one crop. The problem is further exacerbated by the rapid evolution of insecticide resistance in organisms with large populations and short life-cycles, and continuous breeding of ever-more pest- or disease-resistant crops which in turn may stimulate the evolution of more virulent pests and pathogens.

Table 6.5 Percentage of global preharvest losses due to weeds, insects or diseases

	Percentage total loss			Weight (million tonnes)		
	Maize	Wheat	Rice	Maize	Wheat	Rice
Causes of loss						
Weeds	37	40	23			
Insects	36	21	58			
Diseases	27	39	19			
Total loss preharvest (*L*)				121	86	207
Harvested crop (*H*)				128	266	232
$L/(L+H) \times 100$	35.7	24.4	47.1			

(from Spedding, 1975)

It has been estimated that weeds, pests and pathogens may account for a reduction of global preharvest yields by nearly 50 per cent in some crops (see Table 6.5). All depress yield mainly by reducing the LAI of the crop with which they are associated. Weeds also compete with the crop for incident light and nutrients and, particularly in the early growth stages, can reduce yields drastically. Agricultural soils may contain a considerable store of viable weed seeds, some of which were originally sown with the crop; some a residue from a previous crop; others imported from other areas. Cultivation and crop rotation were traditionally the most effective methods of weed control. Greater specialization accompanied by a reduction in the use of rotations together with an increase in continuous arable monoculture has resulted in a greater carryover of weeds, pests and pathogens in arable soils. Poor quality grass-seed has within the last 5–6 years created increased weed problems in Britain. The expansion of winter cereals on to formerly improved grassland has been accompanied by a rapid spread of weeds, particularly of black grass (*Alopecurus myosuriodes*), soft brome (*Bromus mollis*) and wild oats (*Avena fatua*).

Pests and pathogens reduce yield directly by consuming or spoiling the harvestable organ or indirectly by decreasing the size or effectiveness of the leaf area. Control now relies heavily on the use of chemical pesticides and the breeding of disease-resistant cultivars. Both methods have, however, created almost as many problems as they have been designed to solve. The rapid rate of reproduction and evolution of ever-more chemical-resistant pests and pathogens has necessitated the concomitant breeding of even more resistant crops. Pesticides can reduce or completely destroy natural pest predators as well as the target pest organisms themselves. Additionally, some chemicals are more persistent than others and can be taken up from the soil by plants and incorporated in food-chains with deleterious effects on organisms other than the target species. More recently, with increasing knowledge of entomology and the factors controlling insect populations, there has been a revival in the use of biological methods of control. These include encouraging or introducing suitable predator species, and endeavouring to maintain, by careful monitoring and management, a high diversity of insects with low self-regulating population levels, more akin to those in wild ecosystems. The most developed approach to the problem is to use all methods available with varying emphasis for different crops at different stages in their growth cycles. This is the basis of what is known as *integrated pest management.*

WEATHER

Finally, yields vary depending on the *weather conditions* at each growth phase of the particular crop up to, and including, the maturation of the harvestable part of the plant but, more particularly, during the most critical phase. All other factors being satisfied, maximum yield will depend on the occurrence of optimum temperature, light, and soil water levels during each development stage. Depression of yield due to bad weather will depend on when it occurs and how severe the limitations are or by how much and for how long the actual conditions deviate from the optimum. Crops whose economic yields comprise the bulk of the above-ground vegetative production (i.e. green vegetables) are particularly susceptible to water deficits throughout the year. For the cereals the critical phase is just before and during flower formation (anthesis). Water deficiency is critical for soft fruits during the production phase. Suboptimal temperatures retarding growth are most serious in the earlier crop phases; and low temperature combined with high precipitation and low sunshine levels can delay maturation, and reduce the quantity and quality of both cereals and fruit.

Biscoe and Gallagher (1978) have noted that the processes of vegetative growth and grain production are independent in cereals. As a result the effect of weather on dry-matter production *per se* does not seem to exert a direct influence on the rate of grain growth. Except under very high temperatures and severe water stress, mean grain weight is relatively constant and yield is a function more of the number of grains produced per unit area. However, water stress 5 weeks before ear emergence in wheat can result in a 70 per cent decrease in grain yield but only a 52 per cent decrease in total dry-matter production; comparable figures given for maize are 47 per cent and 30 per cent respectively.

Although understanding of the relationships between crop yield and weather has increased considerably in recent decades, efforts to establish a statistical relationship between yield and particular weather components have failed to produce consistent results; and empirical hypotheses are difficult to substantiate in the field. Hudson (1977) has suggested that one of the ways to assess the relative importance

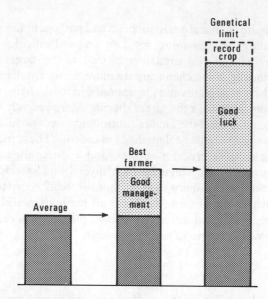

Fig. 6.7 Increases in yield due to good management (from Hudson, 1977)

of the effect of weather is to compare average and record yields in particular localities as shown in Fig. 6.7.

INCREASE IN CROP YIELD

Before the agricultural revolution of the eighteenth century, crop yields everywhere were very low compared with those achieved today. Grain yields of only 0.25 tonne ha^{-1} were recorded in medieval Britain (Evans, 1980). With the enclosure of land and the application of new farming techniques, agricultural production began to rise. New methods of cultivation – particularly crop rotations and the use of the 'grass-break' – helped to depress weed competition while ensuring the build-up and conservation of soil fertility. Towards the latter half of the nineteenth century, traditional nutrient sources such as marl, bonemeal, manure and sewage began to be supplemented by mineral fertilizers. In addition, new varieties, particularly of wheat, better adapted to the newly opened agricultural land in the subhumid climatic areas of America, Eurasia, Australia and South Africa, were being developed. By the end of the century, yields were showing a marked upward turn (see Fig. 6.8). This was seen earliest in Japan, the Netherlands and Denmark, where apart from the drought years there has been a steady increase in cereal yields since the beginning of the century. However, all the agriculturally developed countries shared in the dramatic increases in crop yields following the Second World War consequent upon a high and continuing input of fertilizers, pesticides and insecticides and the development of crop varieties with high yield potentials. It has been estimated that world fertilizer consumption has increased about five times since 1945, and that 36–55 per cent of the present yield of the four main arable crops in Britain (barley, wheat, potatoes,

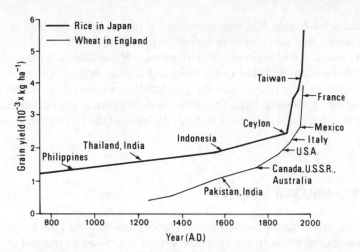

Fig. 6.8 Historical trends in the grain yield of rice in Japan and of wheat in England compared with 1968 yields of rice and wheat in selected countries (from Evans, 1975)

sugar beet) is the result of fertilizer input (Hood, 1982). As Table 6.6 indicates, increase in the use of nitrogen has been most marked and sustained; that of phosphorus and potassium has stabilized at levels necessary to replace loss by cropping.

Table 6.6 Fertilizer usage in the UK

Year (June to May)	Fertilizer usage (k tonne)		
	N	P_2O_3	K_2O
1939–40	61	173	76
1949–50	229	468	238
1959–60	410	458	430
1969–70	796	476	419
1979–80	1268	440	444

(from Hood, 1982)

The production of new higher-yielding cereal varieties was effected by the long-established process of pure-line selection, by which means desirable characteristics are selected from existing varieties, self-fertilized and propagated until a true breeding strain is developed. A major breakthrough was achieved with the production of hybrid maize by cross-fertilization of different varieties. It came into commercial production in the USA in the 1920s and 1930s and by the 1950s had been adopted throughout the Mid-West Corn Belt, where yields increased five-fold in 20 years. Hybrid maize has a very high yield potential, but it does not breed true and its cultivation is dependent on the use of new seed each year, produced by specialized seed-banks. It is also genetically uniform in comparison to the more variable 'land-races' and cereals produced by pure-line selection and consequently it has less resistance to unfavourable environmental conditions. Maize, because of the morphological separation of its male and female organs and natural cross-pollination, was easier to hybridize than the small-flowered self-pollinating cereals such as wheat, barley and rice. The application of hybridization to improved yield potential in these crops came later and was the basis of the Green Revolution of the 1960s.

THE GREEN REVOLUTION

The success following the Second World War of plant-breeding programmes aimed at increasing crop yields in the agriculturally underdeveloped tropical areas of the world gave rise to the term *Green Revolution*. The aim of the programmes was to produce varieties – particularly of wheat and rice – which would be capable of high yields: the HYVs as they became known. The traditional 'unimproved' varieties of these crops were tall, with long lax leaves, a low harvest index and deep widely spreading root systems. Yields were exceptionally low, being limited by poor soil fertility and, more particularly, a deficiency of nitrogen, which is greater in the Tropics than elsewhere. These low yields, however, were to a certain extent compensated by the high genetic diversity of varieties that had evolved by selection over a very long time and which gave an inbuilt resistance to drought or disease. Higher yields could not be attained by increasing planting density because of the lack of nitrogen and the depressive effect of greater leaf-shading, while increased use of fertilizers merely resulted in greater leafiness and mutual shading together with elongated flower-stems which became increasingly susceptible to lodging (Jennings, 1974). The principal outcome of the Green Revolution was to produce dwarf or semi-dwarf varieties of cereal crops with stiff stems and short upright leaves which allowed dense planting, with minimum shading and relatively constricted root systems, and the potential to give high yields when supplied with adequate fertilizers, water, and disease protection (see Fig. 6.9).

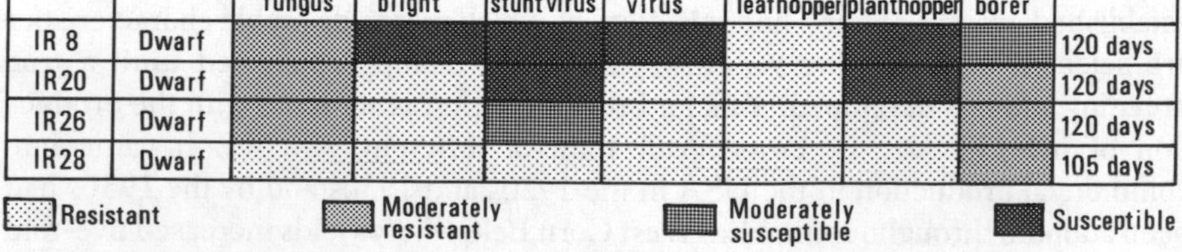

VARIETY	STATURE	DISEASE RESISTANCE				INSECT RESISTANCE			GROWING SEASON
		Blast fungus	Bacterial blight	Grassy stunt virus	Tungro virus	Green leafhopper	Brown planthopper	Stem borer	
IR 8	Dwarf								120 days
IR 20	Dwarf								120 days
IR 26	Dwarf								120 days
IR 28	Dwarf								105 days

Legend: Resistant | Moderately resistant | Moderately susceptible | Susceptible

Fig. 6.9 Characteristics of rice varieties produced by IRRI breeding programmes (illustration by A. Christie from Jennings, 1976 by courtesy of *Scientific American*)

The breeding programme started in Mexico in 1943 with the improvement of spring wheat. Yields in the Tropics were, on average, $750 \, kg \, ha^{-1}$ in 1940; by 1970, $3200 \, kg \, ha^{-1}$ were being produced and seed was exported to India, Pakistan and Turkey. In 1960 the International Rice Research Institute (IRRI) initiated research on rice and in 1966 the very first successful new variety – IR8 – was released for use in the Philippines. It was high-yielding, insensitive to day-length and adaptable to a wide range of environmental conditions. Maturing in 100–120 instead of 160 days, it could produce two crops in one year. Other varieties followed and the need to produce those which combined disease and drought resistance with high yield increased: in gaining the latter, the new genetically uniform strains of rice lacked the resistance to environmental hazards of the traditional varieties. Further, as Jennings (1974) notes, the impact of disease, insects and weeds on agriculture is very much greater in the Tropics than elsewhere. Also, new varieties retain their competitiveness for only about half the time they would in temperate climatic areas. Hence the crop breeders have an even greater struggle in the Tropics than elsewhere to keep pace with fast-evolving pests and pathogenic organisms.

Agricultural research in the developing countries is conducted by a network of international institutes in co-operation with national research programmes (see Table 6.7). Each of the institutes is concerned with particular crops (or livestock) and some of them confine their interest to a particular region or climatic regime. All the crop-research institutes employ an interdisciplinary approach in which plant breeders, plant pathologists, entomologists, economists and others work together to improve the productivity of crops. Since 1971 the institutes have been funded in large measure by the Consultative Group on International Agricultural Research (CGIAR), which is an association of national governments, specialized agencies of the United Nations and private philanthropic foundations.

The products of the Green Revolution did not prove to be the hoped-for panacea to the agricultural problems of the developing countries. Their cooking quality and palatability were not the same as those of traditional rice varieties and they were, as a result, less acceptable to subsistence farmers. Also, the cultivation of the HYVs required high inputs of fertilizers, water, herbicides and pesticides, the cost of which was far beyond the means of the small farmers. Hence the new varieties made a greater impact on the more affluent and larger landowners than on those whose need for food was greater.

MAXIMUM YIELDS

There are indications that some of the highest yielding types of crops are approaching the maximum limits set by biological constraints. Potential yield for a given crop can be estimated, given the annual radiation regime, the percentage of light intercepted by the leaf canopy and the harvest index (see Table 6.8). Average annual *record yields* on experimental farms are indicative of the highest possible yields that have been attained by a given crop in a particular environment under the currently most advanced technology and management. Average farm yields, in both developed and developing countries of the world, still represent only a relatively small

Table 6.7 International agricultural research institutes in developing countries

Institute	Areas of research	Funded location
International Rice Research Institute (IRRI)	rice	1960 Philippines
International Maize and Wheat Improvement Center (CIMMYT)	wheat, maize, barley tritical	1966 Mexico
International Institute of Tropical Agriculture (CIAT)	corn, rice, cow peas, soya beans, lime beans, root and tuber crops	1969 Colombia
International Potato Centre (CIP)	potatoes	1972 Peru
International Crop Research Institute for the Semi-arid Tropics (ICRISAT)	sorghum, millet, chick peas, pigeon peas, ground nuts	1972 India
International Laboratory for Research on Animal Diseases (IRAD)	livestock diseases	1973 Kenya
International Livestock Centre for Africa (ICLA)	African livestock	1974 Ethiopia
International Centre for Agricultural Research in Dry Areas (ICARDA)	wheat, barley, lentils, broad beans, oil-seed, cotton	(Planned) Lebanon

(from Jennings, 1976)

Table 6.8 Actual and potential cereal yields in the UK

	Yield (tonne ha^{-1})	
	Winter wheat	Spring barley
Estimated potential yield (Austin, 1978)	12.9	11.1
Record yields obtained (Hood, 1982)	12.0	10.0
National average yield 1979–80 (HGCA, 1980)	5.2	4.1

(from Reece, 1985)

percentage of record yields. However, in some crops such as temperate cereals, record yields are beginning to approach potential yields (Fig. 6.10). To date, increase in yield potential has been achieved by extending the period over which there is a complete crop canopy in order to increase the amount of radiation intercepted; and by increasing the harvest index. The latter has made a very significant contribution to increased yield of cereals. In some cases an index of 50–60 per cent has been achieved; but the maximum limit cannot exceed 60 per cent because any further reduction in the proportion of assimilating organs would be insufficient to maintain a higher harvest index.

The ultimate biological constraint on increased yield is the photosynthetic efficiency of the crop, i.e. the ratio between the solar energy intercepted and the energy of the dry matter produced. It can be expressed by a number of different parameters: time (annual to daily); insolation (total or visible); output (energy equivalent to dry weight, of the total or part of the plant); the economic/utilizable yield in the case of a crop. Efficiency varies with species. Under optimum environmental conditions of high temperatures and light intensities C4 crops which are adapted to more temperate climatic conditions are more efficient than C3 cultivars. In any species, efficiency can, however, vary during the growth period from *c.* 0.18 per cent to *c.* 2.88 per cent, with low values at the beginning and at the end. High-yielding crops grown under optimum environmental conditions may attain a maximum daily efficiency of 10 per cent. On the basis of maximum daily values crops appear to be more efficient than uncultivated plants. However, a comparison of conversion rates over the available growing season (Table 6.9) reveals that only in exceptional cases do crop efficiencies exceed 2 per cent; at best they are comparable to temperate forest efficiencies in Britain.

However, efficiency in terms of economic yields is very much lower, of the order of 0.3–0.4 per cent. Evans (1975) has pointed out that, while agricultural land represents about 11 per cent of the world's terrestrial surface, harvested products account for less than 1 per cent of the total primary biological productivity. Further, actual efficiency is low compared to estimated values of potential efficiency: annual dry-matter production even for a 'good' crop of grain or sugar is less than 3 per cent while the theoretical maximum photosynthetic efficiency is 18 per cent. These comparatively low efficiencies of primary biological productivity in crops are to a considerable extent the price paid for economic production (Alberda, 1962). They reflect energy losses which are incurred in the process of diverting incoming energy along a few selected food-chains. Many crops are annuals whose growth period is less than the potential growing season available; light energy is 'lost' at the beginning because of an incomplete canopy and low temperatures (and/or water stress) and at the end because of leaf senescence as well as, in certain plants, the use of energy in the process of translocation of material to storage organs. Many crops are grown in supraoptimal (for carbon dioxide availability) light intensities. Many are grown over a much wider range than their wild progenitors and, hence, are more subject to reduction of productivity as a result of water, heat and/or nutrient deficiencies. Alberda (1962) records results of experiments which give growing-season efficiencies, for grass and sugar beet grown with ample water and nutrients, ranging between 5 and 6 per cent of insolation (wavelength 400–700 μ). These are compar-

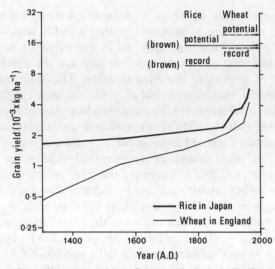

Fig. 6.10 Historical trends in rice yields in Japan and wheat in England. Potential yield ceilings are those estimated for $400 \, cal \, cm^{-2} \, day^{-1}$ during grain filling. The vertical axis is on a logarithmic scale in all cases (from Evans, 1975)

Table 6.9 Comparison of photosynthetic efficiency for types of vegetation and selected cultivated crops

Crop or ecosystem	Location	Growth period (days)	Photosynthetic efficiency (%)
Natural ecosystems			
Tropical rainforest	Ivory Coast	365	0.32
	Denmark	180	
Pine forest	UK	365	1.95
Deciduous forest	UK	180	1.07
Crops			
Sugar cane (March) (C4)	Hawaii	365	1.95
Elephant grass (C4)	Puerto Rico	365	2.66
Maize (two crops) (C4)	Uganda	135+435	2.35
Maize (one crop) (C4)	Kenya (uplands)	240	1.37
Soya beans (two crops) (C3)	Uganda	135+135	0.95
Perennial ryegrass (mean of six C3 cuts)	UK	365	1.43
Rice (C3)	Japan	180	1.93
Winter wheat (C3)	Holland	319	1.30
Spring barley (C3)	UK	152	1.49

(from Cooper, 1975)

able to calculated potential production rates of a close green surface grown under environmental conditions.

MULTIPLE CROPPING

To date, plant-breeding programmes have concentrated on increasing the yield of single-crop stands per unit land area. There are, however, indications that the rate of this process is beginning to slow down. Given the continuing need, particularly in the developing countries, for increased food production, attention has turned to the possibilities of intensifying production by the development of *multiple-cropping systems*.

Multiple cropping involves the production of two or more crops from the same land unit (field) in the course of one growing season (or year, where the two are synonymous), thereby intensifying cropping in both time and space. The crops may be grown simultaneously, i.e. intercropped, or as a sequence of single crops, i.e. sequentially.

The principal cropping patterns within these two systems (Andrews and Kassam, 1976) are:

1. *Sequential cropping*
   (a) Double, triple, quadruple cropping;
   (b) Ratoon cropping, i.e. taking a crop from growth of shoots or seeds after harvest.
2. *Intercropping*
   (a) Mixed intercropping with no distinct row arrangement;
   (b) Row intercropping with one or more crops planted in distinct rows;
   (c) Strip intercropping in different strips wide enough to permit independent cultivation but narrow enough for crops to interact agronomically;
   (d) Relay intercropping during part of life-cycle of each crop; a second crop is planted after the first has reached its reproductive stage of growth but before it is ready for harvest..

Multiple cropping is a very ancient method practised throughout the world and is still the most widespread method of cultivation in the humid tropical countries with a long rainy season and an irrigated agriculture. It varies in form with geographical and energy gradient from very complex intercropping where there is a year-long growing season to sequential and eventually single cropping as the limiting factors of moisture and temperature increase in severity and duration; and from very high labour intensity on small farms (i.e. 5–7 ha) to increasing capital inputs with increasing size, mechanization and crop specialization (Sanchez, 1976). In Zaria Province (North-Central State, Nigeria) Norman *et al* (1984) recorded up to 156 different types of crop mixtures involving two to six crops – those with 2 accounting for 15 per cent; 3 for 42 per cent; 5 for 23 per cent and 12 for 5 per cent respectively of the intercropped area. The most frequent crop mixtures were:

Millet/sorghum
Millet/sorghum/groundnut/cow pea
Millet/sorghum/groundnut
Cotton/cow pea/sweet potato
Cotton/cow pea
Millet/soya bean/cow pea
Soya bean/groundnut

The success of intercropping depends on the ecological compatibility of the crops involved and the extent to which they complement or compete with each other in the use of available environmental resources. Interspecific competition can be minimized by a variety of strategies which include (Harwood and Price, 1976):

1. Using mixtures of similar growth form but of different maturation time (e.g. millet 3 months; sorghum 6 months).
2. Stratification of crops of different potential height as in the case of:
   (a) Annuals planted under tree crops such as coconut, rubber, oil palm etc.;
   (b) Short-season crops which are planted at the start of the growing period of long-duration crops (i.e. corn or soya bean, with sugar cane);
   (c) A mixture of annual crops of differing height, with the tall (corn; cassava) harvested before the short crops (sweet potato); or the short (mung bean) after the taller (corn).

One of the most important advantages of intercropping is the potential for increasing the yield per unit area over that obtained from a single-stand crop grown on the same area (see Table 6.10). It has been estimated that yield can be of the

Table 6.10 Yields of grain by pure and interplanted stands of corn and pigeon peas

Treatment	Grain yield (kg ha^{-1})		
	16 weeks*	24 weeks†	24 week total‡
Corn	3130	–	3130
Pigeon pea	–	1871	1871
Mixed intercrop	2025	1710	3735
Row intercrop	2606	1854	4460

* Corn harvest; † pigeon-pea harvest; ‡ corn and pigeon-pea harvest.

(From Oelsligle *et al*, 1976)

order of 20–50 per cent more when the mixture is one of annual plus perennials. In addition, intermixed crops can provide mutual support (as when vine forms use upright cereals) and protection from exposure to high-intensity rainfall, direct insolation and high wind force. A more complete vegetation cover protects the soil from the accelerated erosion to which all bare mineral soils are susceptible, particularly in the humid Tropics. This also helps to suppress weeds, while the high diversity of crops maintains low pest populations. Also, all types of multiple cropping provide a wide variety of food over a longer period than short-season single cropping and help to reduce the risks of complete harvest failure due to unseasonable weather during the growing season. However, the application of scientific principles and modern agricultural techniques to developing further the production potential of multiple cropping is difficult not only because of the complexity of traditional systems but also because they are based on inherited experience.

The development of early maturing HYVs has begun to increase the flexibility of multiple cropping, particularly of sequential cropping; and in those areas of India where only single cropping could be undertaken this has facilitated double cropping. With a rapidly maturing sorghum replacing the traditional photosensitive *kharif* (wet season) varieties, a second *rabi* (post wet-season or residual moisture) crop can be successfully introduced; or the existing *rabi* crop can be planted earlier and as a result give a higher yield. Multiple cropping was much less frequent in temperate areas of the world where the growing season can be severely limited by low temperatures and frost as well as by low rainfall. In more humid areas, however, it was much more common in the older pre-industrial types of farming when grain mixtures (corn) were often grown. More recently double-row and/or sequential two-crop systems have evolved. The availability of herbicides, the development of shorter season cultivars of small grains and soya bean, and the adoption of minimum- or zero-tillage techniques now allow the speedy establishment and successful maturation of a second crop. This is exemplified by the cultivation of soya beans after wheat or barley in the humid South-East USA (see Table 6.11). Similarly in

Table 6.11 Double-cropping systems practised in South-East USA

Winter crop	Summer crop
Small grains (wheat, barley, oats) for	Corn (grain/silage)
grain-production	Soya beans
silage	Sorghum (grain/silage)
hay	Sorghum – Sudan grass
grazing	Millet
green crop	

(From Lewis and Phillips, 1976)

parts of Western Scotland early potatoes lifted in June are followed by a high-yielding 'catch crop' of rapid-growth hybrid ryegrass which can provide a cut of silage or hay or be grazed until the end of the growing season in September/ October. However, as Oelsligle *et al* (1976) note, it is in the dryland areas of the developing world that the potential for multiple cropping is greatest and most needs to be developed.

AGROFORESTRY

One particular form of multiple cropping is that termed *agroforestry* (MacDonald, 1982). This is a system of land use which combines the cultivation of trees with other crops and/or livestock, either partially or sequentially, and produced both for food and a tree product. The term is relatively recent though the particular type of cropping it defines is very old. In many parts of the developed world, trees are still integral components of the farm systems. The traditional agro-ecosystems of the humid Tropics are frequently characterized by an intimate mixture of trees with perennial and annual crops, all of which provide food and income. The most intensive agroforestry system is that of the Sri Lankan Kandy 'gardens'. These are small farms based on a close association of coconut, kitril and betel palms; with cloves, cinnamon, nutmeg, citrus, durian, mango, rambutan and bread fruit; a lower stratum of bananas and pepper vines; and a peripheral ground layer of maize, cassava, beans and pineapples. As such, this combination provides the farmer with a constant food supply and an income dispersed throughout the year and gives a greater degree of stability than one or two crops. The significance of the tree crop is also related not only to the protection it affords the soil but to the maintenance of the nutrient cycle as a result of a constant supply of tree leaf litter. Indeed a major programme has been initiated in Sri Lanka to establish agroforestry on abandoned tea plantations, which are particularly susceptible to accelerated soil erosion.

Agroforestry has developed from two different forms of land use in the Tropics. One is the widespread, and until recently very stable, system of bush fallow and arable food crops. The stability of this system has been attributed to the deep-rooted trees and shrubs, many of them nitrogen-fixers, which recycle nutrients very efficiently. It is, however, being rapidly undermined by increasing population and a drastic reduction in the length of the fallow period. The second is *taungya* or forest establishment, a form of agriculture the aim of which is to maintain, i.e. conserve, forest and soil in the face of increasingly rapid deforestation. It originated in Burma as a means of using and developing shifting agriculture to re-establish forest fallow by selective tree planting. *Taungya* is now widely used in Ghana and Nigeria to solve the twin problems of land degradation and the high cost of reforestation. As with multiple cropping the ecological benefits of agroforestry include increased use of environmental resources, suppression of weeds, maintenance of soil fertility and stability, and reduction of pest problems. It is, however, debatable as to whether the annual yield of usable products is greater than in non-tree crop mixtures. Other suggested disadvantages are the high degree of interspecific competition between trees and crops and the problem of maintaining a balance between tree and crop components; the difficulties of mechanizing cultivation; and the eventual export of large quantities of nutrients when the system is cropped for wood. There has been a very considerable increase in interest in the ecological characteristics of agroforestry over the last decade. Its potential value as a form of productive land use on all marginal land and particularly as a means of regenerating degraded land in the developing countries has been widely publicized.

Domestic livestock

In comparison to crops, the number of types of fully domesticated animals is relatively small (see Table 7.1). Although domestic livestock are as numerous as humans, they consume four times as much plant matter. However, while animal food (including fish and shellfish) provides only one-tenth of the world's calorific consumption, it is the source of a third of the total protein intake by human beings.

Table 7.1   The main agricultural animals of the world

Animal	*World population 1970–71 (thousands)*	*Number of breeds (approx.)*
Cattle	1 141 215	247
Sheep	1 074 677	230
Pig	667 689	54
Goat	383 025	62
Buffalo	125 412	7
Horse	66 312	124
Ass	41 914	12
Mule	14 733	

(from Spedding, 1975)

The domestic animal (i.e. bred in captivity) has an important role in all agricultural systems. Not least is its ability to convert 'second-class' or low-quality plant protein material into 'first-class' or high-quality animal protein. This is because the herbivorous animal can manufacture its own *lysine*, from lysine-deficient plant material, and produce a meat protein with a high digestibility ratio. The most important domestic animals are cud-chewing *ruminants*, which have the ability to digest plant material high in cellulose, i.e. fresh green leaves and stems. This is a function of the high bacterial content of the enlarged stomach or *rumen* – an organ often likened to a 'living fermentation vat'. As a result the domestic ruminants (including sheep, goat, cattle, yak, water buffalo, camel, llama, alpaca and reindeer) can eat and digest low-quality 'roughage' and produce food from the very high proportion of land that, because of physical limitations of altitude, slope, aridity, cold etc., is not cultivable. Other non-ruminant domestic herbivores (i.e. horse, deer, rabbit) can digest cellulose only with difficulty, while omnivores, like humans and the pig, can obtain little food value directly from green vegetable matter and require more protein-concentrated diets. Domestic animals can also be fed on the natural waste materials of arable agricultural systems or food-processing industries, for instance straw, sugar-beet tops and industrial residues such as brewers' and distillers' grains, sugar-beet pulp and molasses. Finally, the horse (and related species such as the ass, mule and donkey), elephant, camel and water buffalo still retain their value as draught animals.

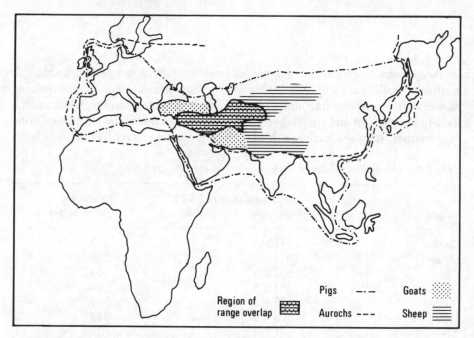

Fig. 7.1 Distribution of wild ancestors of major domestic animals in the Near East (redrawn from Clutton-Brock, 1980)

The Middle East is considered the principal 'hearth' of domestication for all the food-producing livestock (see Fig. 7.1). Their origins are ancient and their evolution seems to have paralleled that of settled agricultural societies and of food crops (Isaac, 1970). The wild progenitors of modern domestic animals were probably common 'scavengers' or 'robbers' of human food. The dog would appear to have been the earliest to make its home with man; evidence exists that it may have originated in the uplands of Northern Iraq *c.* 12 000–14 000 BC, while the cow and the pig appear to have emerged only a little later, *c.* 8000 BC. According to Clutton-Brock (1980) all the common domestic animals were well established as discrete breeding populations, isolated from their wild parent species, by the time of the early

Roman Empire. Zeuner (1963) postulates stages in animal domestication from initial casual contact and/or association with man to farming which facilitated herding and, eventually, confinement and breeding in captivity to produce animals with particularly desirable attributes.

Undoubtedly the possession of certain physiological and/or behavioural characteristics has predisposed (and continues to predispose) to domestication some types of animals better than others (Hafez, 1969). Apart from being an intrinsically useful and easily maintained source of food, good breeding performance combined with a flocking habit making for ease of control and management were obvious initial advantages. Social animals, with behavioural patterns based on a dominance hierarchy, not specialized for instant and rapid flight, have been more readily amenable to dominance by man and have established a high level of communication with him. As in the case of crop plants, so the evolution of domestic animals has been accompanied by physiological changes more adapted to the economic, cultural or aesthetic needs of man and less to the animals' survival in the wild (Bowman, 1977). Among the general effects of domestication have been changes in body size to one either very much larger or smaller than that of the wild ancestors. Selection for features which allow ease of recognition has given rise to considerable diversity in appearance related to skin, hair, and wool colour. Domestic sheep have lost the ability to shed their own wool; and in many domesticates there has been a disproportionate lengthening of the ears or of the tail. Over time, selection has established *breed groups*, each with a uniform appearance that is heritable and distinguishes it from other groups within the same species.

However, until quite recently in this century, the evolution of a breed was a slow process. In part, it was a result of deliberate selection, in part a response to environmental conditions resulting often in geographically isolated groups rather like subspecies, e.g. South Devon cattle, Suffolk Down sheep etc. Increasing specialization, together with artificial insemination in modern intensive farming systems, has allowed cross-breeding on a world-wide scale with the aim of producing animals adapted to the man-made environment of the factory farm.

Sheep and *goats* were the earliest livestock to be domesticated and originated in the highlands of Western Asia. A wide range of environmentally linked breeds developed, particularly in the arid and semi-arid areas of the world. Both sheep and goats have a social system based on a home range, which is usually a restricted area with a *home site* or *core area* within which discrete groups of animals live and graze. Migratory herds often have a distinct winter and summer home range.

Sheep (*Ovis* spp.) comprise eight basic taxonomic groups:

O. aries: domestic sheep
O. nivicola: Siberian snow sheep
O. musimors: European mouflon (Neolithic; the first domestic sheep in Europe)
O. orientalis: Asiatic mouflon: probable ancestor of all domestic sheep
O. dalli: dall or thin-horned sheep
O. canadensis: big-horn sheep
O. ammon: arkar-argali sheep
O. vignen: urial sheep

While sheep are grazing ungulates, originally native to hill regions and mountain foothills, goats (*Capra* spp.) are browsers of shrubs and low trees and tended initially to inhabit the higher bleaker mountain ranges of Europe, Asia and Ethiopia. They did not, however, penetrate as far north as North America. The goat, because of its adaptation to particularly harsh environments, is the most versatile feeder of all the ruminants. An extremely hardy animal, it is capable of breeding and maintaining itself on poor herbage under extremes of temperature and humidity. As a result it has a wider geographical distribution than any other domesticated animal (Harris, 1961).

Today, goats occur in five groups distinguished primarily by horn curvature:

C. hircua: domestic goat (of which there is a tremendous variety)
C. aegagrus: bezoar goat
C. falconeri: markhor from East Kashmir and the Hindu Kush Mountains
C. ibex: adapted to high altitudes: the only wild goat in Europe, found in the Alps
C. cylindricornus: found in the Eastern Caucasus

However, sheep are now commercially more important than goats and are a major source of meat and wool. Goats are usually kept in much smaller groups and used variously for milk, meat, skin and hair. In some parts of the world they are still kept as sacrificial animals and for their supposed medicinal value.

Some two-thirds of the world's goats occur in the Tropics with the largest concentrations in Africa and the Indian subcontinent (Webster and Wilson, 1980), where they are usually dual-purpose. The goat is used primarily for meat and secondarily for milk. Its potential for improving the nutritional standards in the less-developed countries is high. With a gestation period of 5 months, production rate is high. The composition of the milk is nearer to that of human milk than that from the cow and water buffalo and is not so prone to tuberculosis infection as that from tropical dairy cows. Further, the goat can exist on land of limited use for any other type of agricultural production, and thus has been called 'the small farmer's cow'. The ecological role of the goat is, however, debatable. Because of its browsing abilities it has been traditionally cast as the principal factor in the deforestation of Mediterranean lands. On the other hand, it helps to restrict shrub and bush encroachment in areas overgrazed by cattle in the Tropics. Here, according to Webster and Wilson (1980), goats can maintain themselves on land lacking in ground cover and open to erosion as a result of cattle grazing.

The *cow* (*Bos* spp.) supplies two-thirds of the world's animal protein. All modern cattle are descendants of the now-extinct wild ox or *aurochs* (*Bos primogenesis*), formerly ubiquitous in the northern hemisphere of the Old World. Portrayed in the Palaeolithic caves of South-West Europe, it is thought to have had an early religious and social significance, which the modern cow has continued to hold, particularly in the tropical world, where prestige of ownership and economic value are closely linked. Originally a browser and grazer in forests, the cow could also exist on open scrubland. It has given rise to a wide range of geographical races of which the two most important groups today are the European and Asiatic cow respectively (Phillips, 1961). From their original hearth in Western Asia, cattle spread, developing

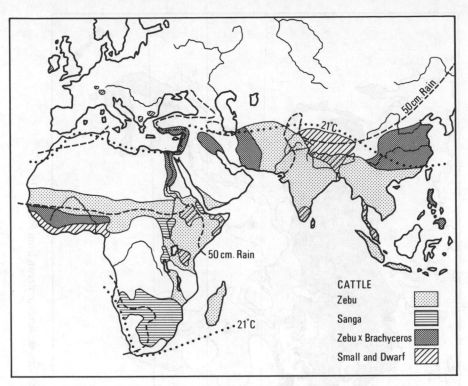

Fig. 7.2 Distribution of cattle types in the Old World (redrawn from Mason, 1984)

local breeds, throughout the Old World (see Figs 7.2 and 7.3). It was not until the fifteenth century that the Saint Domingo breed was introduced to Mexico and hybrids of *B. taurus* and *B. indicus* spread into North and South America. In more recent times breeds of cattle from Europe and the New World have been introduced into the tropical areas of the world, particularly Africa.

European cattle comprise *B. taurus* (humpless), with an early established distinction between *longiformus* (long-horn) and *brachycerus* (short-horn) races. *Asiatic cattle* include *B. namadicus* (the Indian form of aurochs) and *B. indicus*, the domestic humped cattle; *zebu* with a neck hump and *sanga* with a neck and chest hump. The humped cattle are morphologically distinct in having narrower skulls, heavier dewlaps, longer legs, pendulous ears and a muscular or muscolo-adipose hump over the back of the neck and or withers. They are also physiologically better adapted to tropical environments than the humpless animals.

Another domestically important member of the large Bovidae family is the *buffalo* (*B. bubalis*), the domestic water buffalo. With the third largest bovine population in the world it is concentrated particularly in tropical and subtropical Asia. It is a triple-purpose animal, yielding milk with twice the butter-fat content of milk of European dairy cows and good-quality meat when young; and providing a strong heavy draught animal. The *river buffalo* is most important as a dairy animal and a meat producer, particularly in those areas where there is a religious taboo on the slaughter of cows. The *swamp buffalo* is primarily a draught animal used mainly in Malaysia for

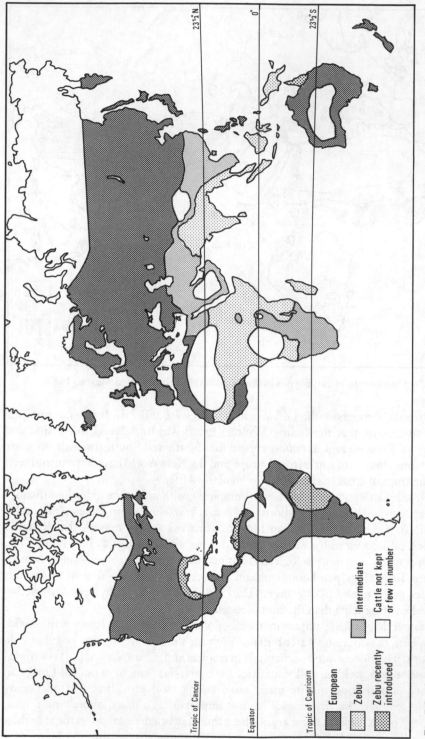

Fig. 7.3 Approximate distribution of the two basic types of cattle and their intermediates (redrawn from Phillips, 1961)

the tillage of paddy fields and for hauling timber. It can digest feed with a higher level of crude fibre (e.g. rice straw) than can European cattle. A recent hybrid, 'cattalo' or 'beefalo', has been produced as a result of crossing *B. taurus* with the undomesticated North American bison (*B. bison*).

The *pig* (*Sus* spp.), descended from the wild boar (*Sus scrofa*), is second only to the cow in number and, together with the dog, had an important initial role as a scavenger of human food. Versatile feeding habits and a high reproductive rate have contributed to the pig's long-held function as 'the poor man's animal', and a most efficient protein producer.

The camel family (*Camellus* spp.) includes three important domesticated or semi-domesticated species (dromedaries, llamas and alpacas), distinguished by their adaptation to harsh climatic conditions of low precipitation and extreme tempera-ture ranges combined with a diet of poor xerophytic vegetation (Wilson, 1984). The most important is the tropical one-humped dromedary (*C. dromedarius*), character-istic particularly of semi-arid regions of Africa. It is capable of maintaining itself on the natural vegetation of shrub and grass as well as on a variety of agricultural byproducts and waste. The dromedary can go for long periods without water and lose up to 40 per cent of its body water before suffering ill-effects. It can also take in and replace lost water very rapidly. Its hairy hide and thick subcutaneous layer of fatty tissue provide efficient insulation in environments where there may be a daily body temperature change of 5.4 °C. The dromedary is a multipurpose animal on which the economy of the desert nomads is dependent. Its milk, rich in protein, fat and vitamin C, is their main staple; its meat is also eaten. Its hide provides material for clothing and shelter, while it can carry people and baggage over very long dis-tances without food or water. The economic significance of the camel has, however, declined very rapidly in recent years with the 'sedentarization' of the desert nomads.

The *horse* (*Equus caballus*) has been the least affected of all the domesticates by genetic manipulation or artificial selection. This may be partly because it is less variable and partly because of its originally specialized function as a draught animal. It shares this function with other members of the Equidae, such as the donkey (*E. africanus*), the ass and the mule.

Trends in modern livestock breeding have been towards a continuing reduction in the number of economically important domestic breeds and for local breeds (or groups) to be replaced by special-purpose livestock derived from a comparatively small number of centres (as in the case of dairy cattle) or by 'custom-produced' hybrids (as in the case of pigs and poultry). Such trends are most advanced in the intensive farming systems of the temperate areas of the world. In the developing countries there is still a greater variety of local types of livestock and of dual-purpose or multipurpose animals, while the use of draught animals for farm work remains important.

LIVESTOCK AND CLIMATE

Climate can affect livestock both directly and indirectly. Temperature is one of the most important direct variables. Domestic animals (including poultry), like humans, are warm-blooded (i.e. *homeotherms*) and hence are most sensitive to the effect of atmospheric variations on their body temperature. An ambient air temperature of over 21 °C causes the body temperature of most European dairy cows to increase; at over 27 °C appetite decline is accompanied by a drop in milk yield. Water consumption in all livestock increases with an increase in temperature. Zebu cattle are certainly more heat-tolerant than European types but whether this is related to body form, body surface area or to the latter combined with a greater density of skin pores is still debatable.

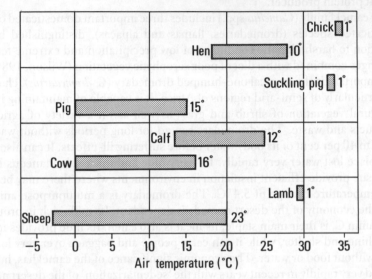

Fig. 7.4 Thermoneutral (comfort) zones and normal body temperatures for adult and infant domestic livestock (redrawn from Cox and Atkin, 1979)

For most livestock the optimal ratio of food intake to weight gain (or to milk/egg production) occurs when the air temperature is 15–18 °C, relative atmospheric humidity is 55–65 per cent, wind speed is 5–8 km h^{-1} and there is a moderate amount of sunshine (McDowall, 1974). As shown in Fig. 7.4 each type of livestock has a *comfort zone* or range of ambient temperature to which it can make homeostatic adjustments and within which it can operate most efficiently with a minimal metabolic rate (Cox and Atkin, 1979). Beyond the limits of the comfort zone, physiological conditions deteriorate. When this happens the animals become more susceptible to other potentially debilitating climatic conditions such as increased atmospheric humidity, solar radiation, wind velocity and precipitation. The thresholds of the comfort zone vary not only with the type of livestock but with the nutritional level and age of the animal (see Table 7.2).

European cattle have a comfort zone between 0 °C and 20 °C and they tend, therefore, to be more cold-tolerant than the zebu, whose comfort zone lies between *c.* 10 °F and 22 °C. Cold tends to increase the expenditure of body energy and to lower survival rates, particularly of young animals. Heat stress, however, combined with high humidity and wind speed is, in general, a more difficult problem, less

Table 7.2 Optimum living conditions for livestock

Type	Temperature range (°C)	Humidity (%)
Calves for breeding	5–20	50–80
Calves while fattening	18–12	50–60
Young breeding cattle	5–20	50–80
Young cattle while fattening	10–20	50–80
Milk-cows	0–15	50–80
Suckling pigs (microclimate for newborn animals	33–22	50–80
Young pigs and pigs for slaughter	22–15	50–80
Sows (pregnant and lactating), boars	5–15	50–80
Lambs	12–16	50–80
Sheep for slaughter or wool	5–15	50–80
Horses (riding, racing or draught)	8–15	50–80
Chicks (microclimate)	34–21	50–70
Female chicks and capons	17–21	50–80
Egg-laying hens	15–22	50–80

Temperature ranges decrease in the case of newborn and very young animals (i.e. calves while fattening; suckling pigs and young pigs for slaughter; chicks). Optimum temperature requirement decreases with age.

(from Seeman *et al.*, 1979)

amenable to amelioration. The type of adaptation to environmental stress also varies. The horse, for instance, can sweat more than either the donkey, the cow or the sheep, which also dissipate body heat by this means. Skin thickness and colour; hair length, colour and oiliness; and the ratio of body surface to mass, all affect heat exchange.

The water requirements of domestic animals vary and some have methods of reducing water loss that adapt them to arid environments. The dromedary is the classic example already noted. It can lose up to 40 per cent of its body water without deleterious effects while maintaining its appetite. In addition, it has a low respiratory rate and a body temperature range of *c.* 5.4 °C, in comparison to 0.6–1.2 °C in most farm animals. According to Mahadevan (1968), zebu-type cattle seem to have lower water requirements than European breeds. Indeed high-producing cattle tend to have a lower potential adaptability to stress than the less-productive primitive breeds. Furthermore the latter can maintain a calving period of 12–13 months compared to 2 years in less-hardy species exposed to similarly harsh environmental conditions.

LIVESTOCK PRODUCTIVITY

Crops may be grown for direct primary use and/or as fodder (feedstuff) for domestic livestock. While the non-utilizable parts of a cash crop, or the waste products of crop processing, are used for animal feeding, the term *fodder crop* usually refers to those cereals, roots, legumes and grasses grown primarily or exclusively for this purpose. Fodder crops are normally harvested and fed to animals, in contrast to *forage crops* such as grasses, legumes, green vegetables and, in some cases, roots, which are grown mainly for grazing *in situ*.

In order to maintain themselves in a healthy condition, all livestock require a basic minimum diet containing the requisite proportions of carbohydrates and/or fats, proteins and minerals (see Fig. 7.5). While most domestic livestock can make

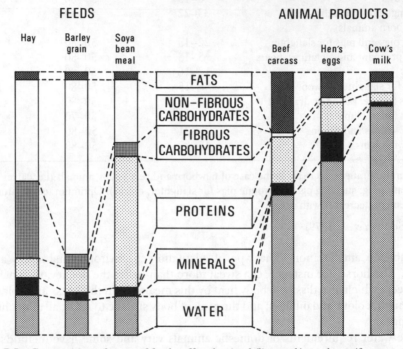

Fig. 7.5 Composition of types of feedstuff and animal (livestock) products (from Greenhalgh, 1977)

use of a wide range of feedstuffs, particular animals differ in their ability to utilize different types of feed; in the efficiency with which they can convert plant food energy and protein into animal protein in the form of meat, milk, eggs or breeding stock; and in the amount of food required for maintenance, growth and reproduction respectively. The use made of forage or feedstuff by domestic animals is a function of its *palatability* and of its *digestibility*, i.e. the amount of food ingested that is retained by the animal. The percentage digestibility or the *digestibility ratio* varies with the composition and, particularly, the nutritive value of the feed; the amount ingested; and with the age and the digestive system of the animal.

With regard to their systems of digestion, domestic livestock fall into two main groups: *ruminants* and *non-ruminants*. *Ruminants* (of which cattle, sheep, goats and deer are the most important), possess a specialized digestive organ, the *rumen*, capable of digesting cellulose and crude fibre (roughage) provided it is not too heavily lignified. The plant proteins in cellulose are resynthesized in the rumen into animal proteins. Digestibility of the feed in this case will be dependent on its C:N ratio. If this is too high, the nutritive value may be insufficient to maintain the overall health of the animal since a minimum amount of nitrogen is required by rumen bacteria to break down cellulose in the rumen (Balch and Reid, 1976). With an increasing intake of concentrates (high-protein feed) the percentage roughage (high-carbon feed) digested tends to fall. While ruminants are adapted to exist on high-cellulose diets, the amount they ingest and can digest is limited by low-quality feed. *Non-ruminants* (of which pigs, poultry, horses and rabbits are the most important), in contrast, are those animals which either cannot digest cellulose or can do so only with difficulty. As with humans, these animals have a specific requirement for each of the nine to ten essential amino acids (see pp. 15–17), which they are unable to synthesize themselves. Diets supplying sufficient amounts of lysine will usually contain the correct amount of the other amino acids needed.

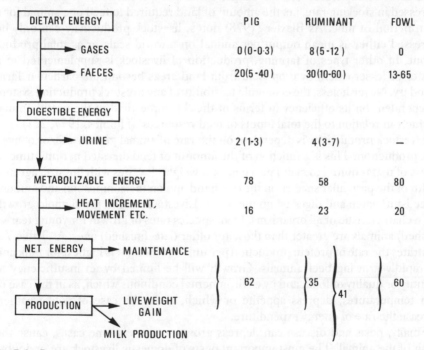

Fig. 7.6 Percentage conversion of gross dietary energy into animal products (adapted from Balch and Reid, 1976)

The products of digestion are absorbed and used to sustain the animal's metabolic processes (see Fig. 7.6). As in plant growth, energy conversion is incomplete: only a relatively small proportion of that digested is retained in the animal's tissues, the remainder being dissipated as heat of respiration. The efficiency of conversion of plant biomass into animal biomass is of vital economic significance since feedstuffs normally comprise the largest single cost in animal production. In its simplest terms *livestock conversion efficiency* (or rate) can be expressed as the ratio of R (resources used, i.e. food intake) to E or P (energy or protein produced) by an animal over a given period of time. Although average conversion rates are frequently quoted for livestock, these are, in fact, highly variable. Among the factors affecting conversion rates are: the amount of feed intake (this is generally determined by appetite, which can vary with environmental conditions, particularly with critical temperatures and, to a certain extent, palatability of the feedstuff); the quality of the feed; the age and health of the animal. While food intake cannot be increased beyond the capacity of the animal, the quality and hence the amount that can be digested can be increased up to a limit determined by the breed and productive capacity of the animal.

In comparison to the primary biological productivity or yield of agroecosystems, secondary livestock productivity cannot be expressed in terms of production per unit land area. In the first place, even on rangeland where livestock are entirely dependent on the available forage, data are not available. Livestock output is recorded in number of animals sold off the range, while the productivity of the range is expressed in stocking rate, i.e. the amount of land required to feed one animal for a given period of time. As Blaxter (1978) notes, livestock productivity can only be expressed either as mean output per animal on a world scale or as total product output. In other types of farming, production of livestock is supplemented to a greater or lesser extent by crop yield from land areas beyond the particular farm boundary. Nevertheless, the economic feasibility of any livestock production system is dependent on its efficiency in terms of the 'life-time yield of the useful animal products in relation to the total inputs of feed resources' (Holmes, 1977, 221).

Livestock productivity is dependent on the rate of animal growth and/or of egg/milk production. This is a function of the amount of feed digested per unit time in excess of that required to maintain a constant weight. It varies with the size, age and health of the particular species on the one hand and on the compostion and amount of feedstuff eaten and digested on the other. Livestock exhibit the sigmoid growth curve characteristic of all organisms. In all species rates of growth for young (early-finished) animals are greater than those for older (late-finished) ones. As Table 7.3 illustrates the rate of protein production per unit body weight is highest for milk and for rapidly growing beef animals. Growth will be limited by an insufficiency in quantity or quality of food; and by environmental conditions which, as in the case of high temperatures, depress appetite or which, as in the case of cold weather, increase the rate of energy expenditure.

Finally, pests and disease can depress growth or, in extreme cases, cause the death of the animal. The most important pests of domestic livestock are undoubtedly parasitic animals. Two main types of pests affect livestock:

1. *Helminths*, worms that inhabit mainly the lungs and gastro-intestinal tract and other organs.
2. *Arthropods*, i.e. insects, ticks and mites that are parasitic on animals.

While relatively few limit production, pests reduce animal health and productivity and make livestock more susceptible to disease. Disease-causing organisms may be

Table 7.3 Typical relative growth rates for meat animals; and typical values for daily formation of protein relative to body weight

	Days to finish	Relative growth rate (g kg^0.75)	Percentage carcass usable*	Protein of g kg^-1	Protein production (g day^-1)	(g day^-1) kg^0.75
Chicken (broiler)	60	31	63	113	2.27	2.3
Eggs	–	–	–	102	4.18	2.5
Turkey (hen)	147	21	79	144	5.85	2.4
Rabbit	60–98	33–22	60	160	3.12	3.1
Pig						
pork	115	39	77	110	45.7	3.3
bacon	165	31	78	105	44.2	2.5
hog	198	26	80	90	37.4	1.9
Sheep						
early lamb	70	47	45	130	25.7	2.8
late lamb	210	18	45	120	11.3	0.9
Cow						
veal calf	131	31	58	140	85.3	2.5
cereal-beef	349	18	54	140	79.4	1.4
18-month beef	569	12	54	140	57.4	0.9
24-month beef	722	9	52	140	47.3	0.7
Red deer	365	11	58	145	12.9	1.0
Milk	–	–	–	33	370.6	3.5

$kg^{0.75}$ metabolic weight.

* The proportion of the live animal that is available for use.

(from Holmes, 1977)

endemic and debilitating and/or epidemic. Recurrent highly infectious diseases which reduce or destroy commercial value and may ultimately be fatal are still among the most severe of agricultural hazards. There are some parts, particularly of the tropical world, where the disease hazard is still such as to make commercial livestock farming difficult, if not impossible. *Rinderpest*, which decimated 2.5 million cattle and 80–90 per cent of the giraffe, eland, wildebeest and antelope in South Africa in the 1890–1905 epidemic, is still a major impediment to sustained development, as are *nagana* (or animal *trypanosomiasis*) in tropical Africa and swine fever. Of all the major livestock diseases *foot-and-mouth* can cause the greatest economic losses in any part of the world. There are also a great variety of invasive or infectious organisms (see Table 7.4), some of which multiply within, some merely spending their development period in, the host animal. Some can exist in a free-living state for a relatively long period of time. Transmission of disease in livestock can, as in humans, be vertical by way of parent's milk, sperm etc. or horizontal by one or more of a variety of external agents. These can involve direct or indirect (i.e. via excretions or secretions) contact between animals; transport by air, water, food, clothing,

Table 7.4 Principal types of infectious agents of disease in domestic livestock

Type of agent	Some important livestock infections with which associated
Viruses and microscopic organisms intermediate between viruses and bacteria	hog cholera, vesicular exanthema, vesicular stomatitis, foot-and-mouth disease, rinderpest, rabies, scrapie, encephalomyelitis, bluetongue, Newcastle disease
Bacteria	anthrax, clostridial infections and intoxications, erysipelas, salmonellosis, coliform infections, vibriosis, listeriosis, pasteurellosis, infectious keratitis, tuberculosis, pseudotuberculosis, Johne's disease, brucellosis, leptospirosis, mastitis complex
Fungi (including mould-like bacteria and yeast-like fungi)	ringworm and other dermatophytoses, aspergillosis, coccidioidomycosis, histoplasmosis
Protozoa	coccidiosis, histomoniasis, trypanosomiasis, piroplasmosis, anaplasmosis, toxoplasmosis
Cestodes (tapeworms)	cysticercosis, gid, hydatid disease, many types of intestinal infections
Trematodes (flukes)	fascioliasis, schistosomiasis, a number of others
Nematodes (roundworms)	hookworm disease, mixed trichostrongyle infections, lungworm diseases, ascariasis, strongyle infections, pinworm infections, many others
Insect larvae	screw-worm infection, ox warbles, other myiases

(from Schwabe, 1980)

equipment etc.; or *vectors*, most of which are living invertebrate animals such as insects, mites, ticks and snails. Where vectors are involved, the spread and intensity of the resulting epidemic is dependent on a complex set of interactions between the agent, the host animal and the environment (see Fig. 7.7). While livestock can be affected by infectious or contagious diseases at any time of the year, some outbreaks are related to the age or physiological condition of the animal (e.g. during pregnancy or lactation). Others are associated with the condition of the livestock as affected by seasonal variation in food supply. Finally, the timing of yet other epidemics is related to the seasonality of the agent; and many pests and pathogens are most active in warm weather.

Animal disease can spread very rapidly over wide areas and can quickly assume the proportion of a regional or national problem. Spread is, in the first instance, facilitated by the movement and contact of stock in its passage through public markets and stockyards. Increasing rapidity of interstate and intercontinental trade has exacerbated the problem and has necessitated slaughter programmes and quarantine restrictions that may operate at every level from the national port to the farm gate.

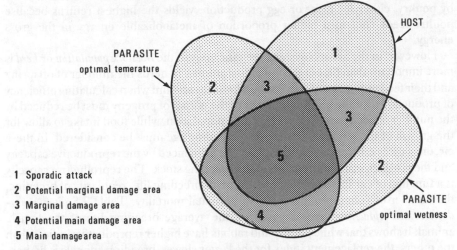

HOST

PARASITE
optimal temerature

1

2   3

3

5

2

1 Sporadic attack
2 Potential marginal damage area
3 Marginal damage area
4 Potential main damage area
5 Main damage area

PARASITE
optimal wetness

4

Fig. 7.7 Model of the effective and potential occurrence of a plant pest or pathogen, controlled by temperature and wetness (from Klinghauf, 1981)

EFFICIENCY OF FEED CONVERSION

The efficiency of feed conversion by livestock (i.e. the ratio of feed input to food output) can be expressed by a number of different parameters including:

1. Protein value.
2. Metabolizable energy.
3. Gross energy.

Since protein is the main product and energy supply is the principal limiting factor (Holmes, 1977), efficiency can also be expressed as the output of protein per megajoule metabolizable or gross energy of the feed input. As indicated in Table 7.5 the most efficient energy and protein conversion is effected by milk production and by meat from pigs and young beef cattle. Efficiency, however, will also be influenced by the amount and composition of the edible food produced by the carcass. The *carcass* (or killing-out) percentage is that proportion of the live animal that is available for use. As can be noted in Table 7.3, the carcass percentage is particularly high for the pig, followed by poultry and rabbits; it is considerably less for the larger heavier sheep and cows. Also, poultry and rabbits produce meat of high-quality protein and with a fat content of less than 10 per cent; in the others, particularly the pig, fat accounts for 25 per cent or more of the edible carcass.

The proportion of the carcass that is edible, however, varies in time and place with culture, diet and taste – with changing market demands for products of a particular quality, form and/or shape. Demand for lean meat has increased the production of young rather than mature animals, while small cuts and joints are preferred to large ones, particularly for the packaged supermarket product. Egg, milk, pig-meat and broiler production also show high efficiencies of conversion of both metabolizable and gross energy. The production of protein per unit of energy by poultry, either for meat or egg production, yields the highest returns because poultry feeds contain a higher proportion of metabolizable energy in the gross energy.

However, in the agroecosystem, the efficiency of the *livestock population* or *herd* is more important than that of the individual animal. In herds the number of offspring and their feed requirements have to be taken into account when calculating efficiency of production. For the self-replacing herd the output of progeny must be reduced by the number needed to maintain a breeding population, while food intake to allow for the growth of the remaining progeny to breeding age must be considered. In these circumstances efficiency will, in addition, be influenced by the reproductive capacity and the replacement requirements of the breeding stock. The reproductive capacity is a function of earliness of breeding, length of breeding life, regularity of breeding, the number of young produced and of perinatal mortality. Table 7.5 gives *reproductive* and *replacement indices* based on the average breeding life of the female animal. It shows that while poultry and rabbits have higher reproductive indices than the others, the replacement index for the larger slower-breeding animals is higher.

The differences between individual animal (Table 7.6) and population efficiencies are relatively small for milk and poultry (whether for meat or eggs). This is

Table 7.5 Reproductive and replacement indices for domestic livestock

	Typical breeding life (years)	Number progeny per annum	Reproductive index (per kg)	(per kg 0.75)	Replacement index*
Chicken	1	100	2.0	2.5	0.01
		150	3.0	3.77	–
		250	5.0	6.28	
Turkey	1.5	50	0.5	0.89	0.013
		80	0.8	1.42	–
Rabbit	1.5	20	0.22	0.32	0.033
		40	0.44	0.65	0.0165
Sow	2.5	12.0	0.11	0.38	0.033
		24.0	0.17	0.61	0.017
Ewe	4.0	1.0	0.06	0.19	0.25
		2.0	0.12	0.33	0.125
		3.0	0.13	0.37	–
Cow	5.0 (suckler)	1.0	0.08	0.40	0.2
	4.0 (milk)	2.0	0.14	0.66	0.25

* Replacement index calculated as female replacements divided by total progeny.
† reproductive index = annual mass of live young born or hatched relative to mass of dam; kg 0.75 metabolic body size.

(from Holmes, 1977)

Table 7.6 Efficiency of single animals from birth, or from the beginning of laying or lactation

	Edible protein (%)*	Edible energy (%)†	Edible energy (%)‡	g protein/ MJ total ME	g protein/ MJ total GE
Broiler	19.0	16.0	11.0	2.9	2.1
Turkey					
stag	20.0	9.0	6.0	3.1	2.2
hen	20.0	9.0	6.0	3.1	2.2
Rabbit	17.0	13.0	8.0	3.2	1.9
Pig					
pork	27.0	31.0	20.0	3.4	2.2
bacon	22.0	25.0	17.0	2.6	1.8
heavy hog	15.0	22.0	15.0	1.7	1.1
Lamb					
early	28.0	28.0	25.0	3.3	3.0
late	10.0	15.0	9.0	1.3	0.8
Cow	25.0	12.0	10.0	2.9	2.2
cereal-beef	12.0	10.0	6.0	1.3	0.8
18-month beef	11.0	10.0	6.0	1.3	0.7
24-month beef	9.0	10.0	6.0	1.1	0.6
Milk					
low conc	20.0	21.0	11.0	2.5	1.4
high conc	21.0	23.0	13.0	2.8	1.5
Chicken	25.0	21.0	14.0	3.2	2.1

* Percentage total crude protein eaten.
† Percentage total metabolizable energy (ME) eaten.
‡ Percentage total gross energy (GE) eaten.

(from Holmes, 1977)

because the dairy cow has a relatively long lactation period (4–5 years), while the hen starts to lay at 3–4 months old, and can reach maximum weight in *c.* 10 weeks as well as producing a large number of progeny. Both require high-energy feed for production. Meat production is generally less efficient and the difference between that of the individual and of the population is greater; efficiency of the herd can be reduced to half or less than that for the single animal (Holmes, 1977), This is because replacement rates are high and reproductive indices are low. Of the three meat producers, the pig with a very high reproductive rate is the most efficient; the sheep and suckler cow the least. It must, however, be remembered that the commercial production of poultry and pigs is based almost entirely on concentrated feed, while cattle and sheep can utilize cellulose. While milk production and cereal-

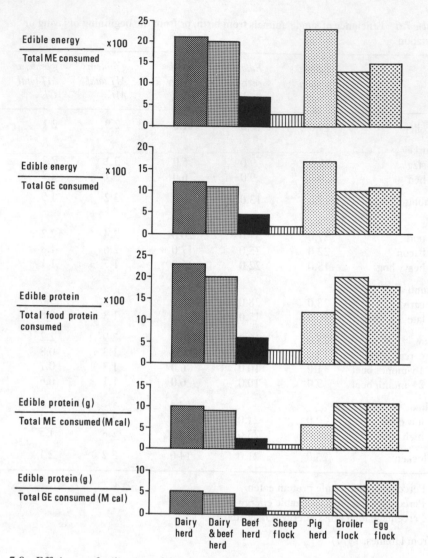

Fig. 7.8 Efficiency of utilization of dietary energy and protein in whole-farm situations. Values are shown for five expressions of efficiency in seven farming systems. ME = metabolizable; GE = gross energy (from Balch and Reid, 1976)

beef obtain 50 per cent and 95 per cent respectively of their metabolizable energy from concentrates, British cattle in general obtain 20–40 per cent and sheep only 6 per cent of their energy from this source. Finally, the efficiency of livestock production in livestock farms is illustrated in Fig. 7.8. In terms of energy conversion, dairy, dairy–beef and pig herds are the most efficient; beef and sheep the least. Dairy, dairy–beef and poultry are the most efficient converters of plant to animal protein, while egg- and meat-producing fowl are the most efficient livestock in terms of the conversion of plant energy into edible animal protein.

As with crops, livestock productivity has increased dramatically since the 1940s, as reflected in the number of eggs per laying hen; in the milk yield per dairy cow; and in the meat production per animal or herd. Livestock efficiency has been improved, particularly in Western agriculturally developed countries, by the breeding of livestock with high reproduction rates and rapid growth or production potential combined with the use of protein-rich (concentrated) feedstuff. The latter accelerates growth rates and, as a result, reduces the final carcass mass. While there are indications that egg, milk and pig production may have attained their maximum potential efficiency, the last tends to be depressed by neonatal mortality. Continuing efforts to increase livestock efficiency have focused on increasing the growth rate of male animals consistent with market demands for size and composition of meat cuts and by increasing the fecundity of animals such as sheep with a low reproductive rate. Alternatively, it has been suggested that livestock productivity could be further improved by using more productive feed crops, increasing the productivity of traditional feed crops, and/or by making better use of crop residues and byproducts. Livestock efficiency in the less-developed countries is still extremely low. It is difficult to see how the combined constraints on productivity of social custom, a genetically poor stock, the lack of concentrated feedstuff and low levels of animal nutrition together with a high level of pest and disease can be easily overcome in either dairy or beef cattle, pigs or fowl. Also, increasing population puts such a high demand on food production that animals must continue to be supported by the poorer land unsuitable for cultivation. Under these circumstances the goat may have the greatest potential for increased milk and meat production.

8

Land capability for agriculture

Land is a basic but finite resource. It still supplies most of man's food, as well as his living and working space. Ownership of or any other territorial rights to land are long-established bases of wealth, status and power in human societies. Land is the major component in all but a few specialized types of agriculture. In agricultural terms, at least, the concept of land includes 'the soil together with all other bio-physical attributes of the environment (e.g. climate, relief, soil parent material, organisms) which affect production therefrom' (Vink, 1975).

Recognition of the variation in value of agricultural land is reflected in attempts to reach an equitable distribution of good, medium and poor land in traditional communal farming systems in Europe and Asia. Categorization of land for the purpose of settlement and taxation go back a long way in time. That based on 'land that can feed a person' was used by the Romans; and a similar system was common under Islamic rule in Asia, especially India. Attempts to classify land in terms of its economic value developed in Europe in the early eighteenth century. The Milanese cadastre, whereby every parcel of land was valued, was introduced in Lombardy in 1718–60. Later, in 1864, the Prussian government established a system of equating land prices and soil types which could be used to calculate average purchasing prices and appropriate land tax levies.

However, until the 1940s most systems of land evaluation were based on actual use and production under prevailing economic conditions. As the demands for land for purposes other than agriculture (e.g. housing, transport, industry, recreation, forestry etc.) increased, so its value (unless specifically zoned) was determined more by the competitiveness of other users than by its biological productivity. In order to overcome the problems inherent in economic valuation, attention was drawn to the idea of evaluating land on the basis of those relatively constant biophysical attributes or characteristics which determine its *capability* or inherent potential for agricultural production, i.e. its physical capacity to produce a given crop (output) for an indefinite period without exhaustion, waste or degradation. In some instances a distinction is now made between the concepts of *capability* and *suitability*; in others the two terms are either used synonymously or suitability has superseded capability. The FAO (1978) defines *suitability* in terms of the fitness of a given type of land for a particular purpose which embodies the concept of use on a sustained basis.

There have been three main approaches to the assessment of the capability or suitability of land for agricultural use: the empirical, the parametric and the approach based on limiting factors or conditions.

EMPIRICAL APPROACH

The empirical approach is based on observed land features, often combined with visible indications of performance (e.g. condition of crops and land, size and condition of associated farm buildings etc.), and crop yield. Such assessments are very much dependent on the knowledge, skill and experience of the assessors and must have a high degree of inbuilt subjectivity. It is not always easy on the basis of observation alone to distinguish between the effect of management and of land quality; and the rating may reflect actual rather than potential use or value. However, this empirical method has been and can be used where quantitative data are not available but a general classification is required for immediate planning decisions on a regional or national basis. Stamp's (1947) classification, drawn up immediately after the Second World War, was based mainly on existing land uses. The Department of Agriculture, Food and Fisheries (Scotland) still rates agricultural land according to seven categories, from very good (A+) to poor (C) and non-arable, on the basis of visual inspection combined with average regional crop yields.

PARAMETRIC APPROACH

The parametric approach is dependent on the use of 'key' land qualities or functions that have been identified by significant statistical correlations between them and crop performance and yield. Land areas are rated according to a particular combination of *desirable* agronomic qualities. The latter may or may not be weighted to allow for relative importance, and be summed or multiplied to give a final numerical score which is the basis of high to low capability ratings. The advantages of this method are standardization and accurate replicability. The main disadvantage is the need for, but lack of, comprehensive data on land qualities. Further, there is difficulty in expressing many qualities in a simple quantitative form and numerical ratings often give a spurious impression of accuracy to what are still relative values.

Early examples of this parametric method are the *Storie index* (Storie, 1954) and its derivatives developed in the USA, together with various land-evaluation schemes used in Eastern Europe prior to the Second World War. The Storie index is claimed to give quantitative expression to which particular soil possesses conditions favourable for crop production. It is an inductive method which rates land on the basis of what are considered to be the most important soil characteristics which determine the potential productivity of the land. The characteristics most commonly used for rating are:

1. General character of the soil profile, particularly stratification and degree of weathering; texture and structure; inherent fertility.
2. Topography.
3. Soil 'modifying' conditions such as drainage, acidity, alkalinity, erosion etc.

Each land quality is given a numerical value, the sum of which is expressed as a percentage of the maximum possible. The rating index is calculated by multiplying the point scores for the three main parameters and expressing the total as a percentage of the maximum possible, e.g. $\frac{A \times B \times C}{300} \times 100$. A variant developed in Canada is illustrated in Table 8.1; it was considered that an index of 30 would be

Table 8.1 Canadian variation of the Storie index used to assess land capability for spring wheat in Saskatchewan

Land qualities	Points
A. Soil profile	
texture	40
structure	30
natural fertility	30
B. Topography	100
C. Modifying factors	
climate	25
salinity and alkalinity	25
stoniness	25
tendency to drift	25

(simplified from Smith and Atkinson, 1975)

indicative of conditions marginal for arable land on the basis of four factors (climate, soil, soil depth and water). German soil ratings (*Bodenboutierung*) also exemplify the application of the parametric approach for the purpose of a rapid re-assessment of taxable land values after the Second World War. A numerical rating of yield capacity (i.e. capability) of land as determined by natural factors (i.e. soil, climate, topography and water); market and labour conditions; and nature of the holding was calculated in order to achieve a relatively uniform standard of rating. A national standard estate was given a rating of 100 to which all other actual ratings could be related.

Some parametric methods of land rating have been based mainly on the factors assumed to determine crop yield. Others are based on yield and quality of crop production under physically defined systems of management. The latter, known as *productivity ratings*, are quantitative expressions of the comparative productivity of different soil types. They may be calculated under 'average' and 'good' management. In the USA, 'standard yields' are averages without the use of fertilizer and are selected to represent the approximate average yield obtained for a particular crop in a major soil region in which the crop is the principal agricultural product. For example, Marbut's *productivity index* is derived from the formula:

$$100 \left(\frac{Y_1}{Y_0} \times \text{pr} \right)$$

when Y_1 = yield of given crop on a particular soil type; Y_0 = yield of the same crop under the most favourable conditions; and pr = the percentage total area of a particular soil type occupied by a given product. The difference between productivity ratings, then, is a measure of the soil's response to good management. This method, however, is based on the assumption that the variety of crops grown will everywhere be that which produces the highest yields under the given conditions.

LIMITING FACTORS OR CONDITIONS

This is now the more widely used approach to the classification of land capability. It is based on the absence or lack, rather than the presence, of those land qualities which determine agricultural potential. In this case, land areas are rated according to the particular type and severity of the main inherent physically limiting conditions of local climate, soil and relief, given a level of management and constant regional climatic and economic conditions. Limitations economically capable of rectification, such as nutrient deficiencies, drainage, flood control or irrigation needs, are not taken into consideration: the land is classified as though these had been overcome. In various schemes hazards to cultivation are recognized and defined as limitations which, if ignored, will result in permanent deterioration of the land resource. They can include erosion, flooding, salinity and pollution. In capability ratings hazards may be considered with or independently of the other limitations.

Three levels or scales of capability classification are usually identified (see Table 8.2):

Table 8.2 Tentative land – capability classification in a given area

Soil mapping unit (capability unit)	Capability class	Sub-class*
A	III	s/g
A (steep slopes)	VI/III	g
B	II	s
C	I	s
D	III/IV	w/s/g
E	III	s
F	III	s

* s = soil; g = gradient; w = wetness.

Classes: land areas ranked according to severity of limitations from that with highest potential for arable use to that with no potential.

Subclasses: land areas identified by the major kind(s) of limitation to which a particular area is subject.

Units: soil units or groups of soil units sufficiently uniform to grow similar crops, to require similar management and to be similarly productive.

The concept and application of this approach to land-capability assessments for agriculture was developed by the US Department of Agriculture Soil Conservation Service in the 1930s in response to the need for a systematic interpretation of land characteristics for agricultural advisory purposes and particularly for the implementation of soil-conservation measures following the erosion problems of the late 1920s and early 1930s.

The US classification recognizes eight capability classes based on limitations for arable cultivation and increasing risk of soil erosion:

I   Very high capability
II   High capability
III   Moderate capability
IV   Low capability
V   Marginal for cultivation
VI   ⎫
VII ⎬   Not capable of arable agriculture
VIII   No commercial use for agriculture

Subclasses are defined on the basis of five kinds of limitation: erodibility, wetness, soil (root zone), topography, and climate (see Table 8.3).

While soil conservation is still a primary aim in land-capability assessment in the USA and elsewhere, the concept has widened considerably since its introduction. From the late 1940s it became generally accepted that the agricultural evaluation of land should be based on natural or inherent characteristics rather than on management skills or prevailing economic conditions. While the US system of land-capability classification provided the blueprint, it was adapted to suit the particular needs and problems in other developed and developing countries of the world. Many now have their own national, state or regional schemes (see Table 8.4). The number of classes identified, however, varies, as do the number and type of limitation recognized. The UK systems, for instance, put more emphasis on slope and soil limitations and less on soil erosion, which until relatively recently has not been considered a major agricultural hazard. The Netherlands is more concerned with the evaluation of land for arable and improved grassland than for rough grazing or woodland; and the Canadian Land Inventory separates capability for agriculture from that for other land uses such as forestry, wildlife conservation and recreation. The land-capability concept can be, and has already been, applied to specialized types of agricultural production or types of land management such as irrigation, drainage etc.

The actual assessment and mapping of land capability raises many problems and has, partly as a result, been subject to considerable criticism. The first problem is that of defining the degree of severity of the particular limiting factors used in such a way that identification and classification can be replicated and standardized. Some limitations are described qualitatively, others quantitatively, depending on the data available, the nature of the attribute and the scale at which the classification is being used. At a national or regional level, existing data has to be used or adopted; at the farm level, field measurements prior to assessment may be necessary and economically feasible. Agriculturally significant climatic characteristics such as length of growing season, rainfall amount and duration, frost and drought hazards, and soil water surpluses and deficits can be obtained or derived from long-term meteorological records available in developed countries. The density and distribution of recording stations, however, is rarely such as to provide all the data required to assess limitations at a local or microclimatic scale. In the UK there are few first-class meteorological stations above 300 m, at which altitude the deterioration in climatic conditions – particularly the decrease in length of the growing season, increase in rainfall, decrease in evapotranspiration and increase in exposure to high wind force – is very rapid indeed. The effect of relief on temperature conditions can be based on the normal lapse rate, i.e. a decrease of 1 °C for every 300 m increase in altitude. Increasing severity of exposure, however, is much less easy to assess and express. It can be derived from the 'tattering rate' of standardized flags under varying recorded wind speeds. However, actual measurements are limited in time and duration and have usually been taken to assess the effect of exposure at high altitude on forest rather than agricultural productivity.

Limitations associated with the soil root zone tend to dominate all capability schemes. This not only reflects the significance of soil characteristics in the agricultural evaluation of land but also the initial derivation of land-capability classifications from pre-existing soil surveys in the USA, the UK and in many other countries. Of the three most commonly used attributes, depth and stoniness (at the surface or in depth) can be expressed quantitatively, the former in absolute, the latter in relative terms (i.e. percentage area or bulk occupied by mineral fragments over 2 mm diameter). Soil water limitations, however, are more usually expressed in terms of soil drainage characteristics, as determined by texture, from freely to imperfectly or poorly drained, or in terms of climatically controlled water surpluses or deficits.

The final group of land attributes are those related to the form of the land surface. Of these, surface slope is the dominant factor limiting agricultural use – directly, through its effect on the ease and safety of all cultivation processes, and indirectly, as an important factor affecting surface-water run-off, infiltration and evaporation and the susceptibility of cultivated soils to accelerated erosion. While slope or gradient can be expressed in absolute terms (see Table 8.5), the soil erosion hazard is a much more complex limiting condition.

Initially erosion hazards were estimated on the basis of gulley density, which was a description of the existing severity of erosion rather than an assessment of vulnerability to erosion. Susceptibility of land to *accelerated soil erosion* is dependent on the interaction between climatic conditions, soil type, land form and plant cover. Commonly used measures of the erosion hazard are the *erosivity index*, the *erodibility index* or a combination of the two. Erosivity is a function of the intensity or effectiveness of the eroding agent of rainfall or wind. Erodibility, however, is an expression of the resistance of soil to both detachment and transport (Morgan, 1980), the two main agents of which are water and wind. Resistance to both depends on very similar soil properties except that dry soils are more vulnerable than wet soils

Table 8.3 Land capability classification of US Department of Agriculture Soil Conservation Service

Class	Capability		Subclass	
			category	degree of limitation
I	No or few limitations, level, deep, well drained, friable	may be used safely for all cultivated crops, pastures and forests	Ic	no limitations / less than ideal
II	Generally good soils, with depth, structural features and drainage features only a little less than ideal	may be used for crops, pasture and forests but needs appropriate management	IIc IIs IIw IIe	slight
III	Soils on moderately steep slopes or with impermeable subsoil, or shallow or liable to flooding	may be used for a restricted range of crops, pastures and trees with special management	IIIc IIIs IIIw IIIc,t IIIe IIIs,t IIIs,t,e	moderate
IV	Soils with steep slopes or wind-erosion hazard: shallow, impermeable or low water-holding capacity or salty	may be used for cultivation at considerable difficulty and cost if risk is to be avoided	IVc IVs IVw IVt IVe IVs,t IVs,t,e	considerable

V	Very stony soils, or very prone to flooding or with severe climate	land cannot be cultivated, but for reasons other than erodibility	Vc Vs Vw	fairly severe
VI	Soils with steep slopes, shallow, stony, too wet or with tight subsoils	land unsuited to cultivation but may be used for pastures, trees or wildlife	VIc VIs VIw VIe VIt VIs,t VIIs,t,e	severe
VII	Soils with very steep slopes or very shallow, or salty or too difficult to drain	land unsuited to commercial plants; suitable for natural pastures, forests and wildlife	VIIc VIIs VIIw VIIe VIIt VIIs,t VIIs,t,e	very severe
VIII	Severe slopes, rock outcrops, exposed peaks, sandy beaches, severe climate	land unsuited to commercial plant production; may be used for recreation, wildlife, water supply and aesthetic reasons	VIIIc VIIIs VIIw VIIIt VIIIe	extremely severe

* Subclasses are defined according to the following types of limitation: climate (c); soil (s); wetness (w); erodibility (e); topography (t).

(from Klingebiel and Montgomery, 1961)

Table 8.4 Land capability classification for agriculture by the Soil Survey of England and Wales

Class	Degree of limitation	Principal agricultural crops	Comment
1.	Very minor	all usual British crops + horticultural produce	high yields obtainable: ground suitable for most enterprises
2.	Minor	all usual British crops; increased risk of failure jeopardizes many horticultural crops	high yields obtainable with good management; suitable for most enterprises
3.	Moderate	cereals (all types); grass; forage crops (turnips, kale etc.); potatoes	yields equal to those of Class 1 may be obtained from selected crops under good management; ground suitable for most enterprises
4.	Moderately severe	grass dominant; restricted cereals (mainly oats; some barley)	high yields of a very restricted range of crops possible under good management; suitable for forestry and recreation
5.	Severe	improved grass is the only crop apart from the occasional break-crop	high yield of grass products possible; but risk of failure high; suitable for forestry and recreation
6.	Very severe	no cropping; not improvable except by aerial spray	extensive stock ranching (sheep, deer, cattle); care needed in grazing management to prevent sward deterioration; forestry in parts; some recreational pursuits
7.	Extremely severe	no cropping	Some grazing by hardier stock (sheep, deer) but season restricted to less than 5 months; no forestry possible; some recreational pursuits

(from Bibby and Mackney, 1969)

to wind erosion. Although land form and cultivation are important controlling factors, soil characteristics such as texture, structure, consistence and organic matter content are the most important variables. Erodibility is commonly expressed in terms of soil texture because of the high degree of dependence of the other soil

Table 8.5 Slope classes and tractor–implement combinations

Maximum overall gradient	Slope class	Implement limitations
3°	gentle	no limitations
7°	moderate	limit three-in-line forage-harvesting equipment
11°	strong	combine and trailer equipment with two-wheel-drive tractors
15°	very strong	two-wheel-drive tractors with fully mounted equipment
25°	steep	four-wheel-drive tractor with trailer equipment
30°	very steep	can be used with four-wheel-drive with fully mounted equipment
> 30°	excessively steep	no working possible

(from Bibby, 1982, reproduced by permission of the Soil Survey and Land Research Centre and the Macaulay Land Use Research Institute)

characteristics on this basic and relatively constant attribute. Silty and fine sandy soils are less resistant to water erosion; the most vulnerable are those with a silt content of 40–60 per cent. Some workers such as Evans (1980) prefer to express erodibility in terms of clay content: soils with 9–30 per cent are most vulnerable because of the role of clay in determining the formation of stable structural aggregates and the behaviour of the soil under stress (see Chapter 4).

Simple indices of erodibility have been based either on soil properties or on the way the soil responds to rainfall and run-off and to wind respectively. An example of the former is the *aggregate stability index*, which expresses soil resistance in terms of the percentage of water-stable non-primary aggregates over 0.5 mm in the soil. Another is the K-index which is a measure of soil loss per unit EI30 recorded on a standard bare plot 22 m long and with a gradient of 5° (Morgan, 1980). EI30 is an expression of the kinetic energy in $Jm^{-2} mmh^{-1}$ for a maximum 30-min intensity of rainfall.

Another problem associated with the use of limiting factors in land-capability assessments is that of defining class boundaries. The range of values and particularly the lowest levels considered necessary or acceptable for a particular class may be decided by an arbitrary systematic ranking of values or by reference to known or assumed agriculturally significant limits. In the UK classification, for instance, wetness and stoniness are described qualitatively but can be related to the range of values used in soil surveying. Slope (or gradient) classes are based on the implications for mechanized farming (see Table 8.5). Regional macroclimatic conditions are defined on the basis of the water balance and temperature during the growing season (April to September); local or mesoclimatic classes are zoned by altitude and rainfall (Table 8.6).

Table 8.6   Altitude, rainfall zones and agricultural land use in the UK

Altitude and mean annual rainfall (mar)	Land use
Land >600 m	above tree-line: poor rough grazing only
Land 300–600 m with >150 cm mar	rough grazing; pasture improvement usually not feasible
Land 200–300 m with >125 cm mar	improved pasture but also suitable for arable crops
Land 100–200 m with >100 cm mar	mainly suitable for improved grass and limited arable crops

(from Bibby and Mackney, 1969)

Finally, the number of land-capability classes needs to be related to the range of land conditions within the area to be considered. Where variation is low, fewer classes may be required than where variation is high. In those systems in which there are only four or five classes of cultivable land recognized, the range of variation may be so wide as to make ranking meaningless. In Britain, for instance, a very high proportion of arable land falls into Class I, with the result that the differences in land capability within the class could be as great or greater than those between Class 3 and 2 or between Class 3 and 4.

The US land-capability classification and those derived from it have, since their inception, been subjected to considerable criticism. The system has been con-demned for being too generalized on the one hand and for not being universally applicable on the other. It does not give any indication of the suitability of land for particular crops, farming systems or types of management. It fails to provide an index of real land values or of economic viability. Also, without some sort of productivity rating, it is an inadequate basis for objective land-use planning. In this respect Patterson and MacIntosh (1976) have stressed the importance of relating land capability to production in economic terms. They maintain that land quality, defined by a *productivity index*, is significantly correlated with gross farm returns and gross margin per hectare. It has been demonstrated that high gross returns per hectare are three times as likely with a productivity index of 900–100 than with that of 80–89; and the probability of obtaining high returns at low cost are greatest when land with the highest index is used for production.

This US land-capability classification has, however, been criticized more for what it cannot do – indeed did not set out to do – rather than for its initial objectives. The flexibility and robustness of the concept has been under-rated or neglected. More recently the FAO (1976) has developed a framework for land evaluation guidelines for a standardized land capability [here synonymous with suitability] assessment applicable to any part of the world on any scale. The concept has also been adopted to assess land suitability for particular crops and for soil management. In both cases this requires more precise data than are often readily available about the land conditions required for maximum yield and which will limit production of a particular crop (or variety thereof); or for successful (i.e. economically feasible) management schemes such as irrigation, drainage, zero cultivation etc. Rudeforth (1975) has suggested that land suitability for specific crops could be assessed by comparing sites in which a crop is not presently grown with the range on which it is cultivated. From a selection of sites the range (i.e. the maximum and minimum) of physical conditions under which a particular crop can be grown could be established. All potential sites with any properties beyond the established range would be excluded. The remainder would then be classifiable as land capable of growing the particular crop successfully. However, as the author points out, the method depends on a considerable amount of fieldwork and on a sufficiency of sites on which the particular crop is grown at the time of survey.

The value of land-capability classifications in developed countries with long-established crop and land-use patterns and a high level of management has been questioned. However, as Wilkinson (1968) pointed out, farmers do not necessarily know the full potential of their land. Their perception of its value is often based on parochial standards and can be influenced by existing land use and productivity. He argues that changing farm ownership plus the need to increase economic efficiency have increased the value of land-capability classifications in Britain as a means of providing the basic physical information about a farm's land resources in a systematic and factually acceptable form. His argument can be strengthened by the increasing pressures on land, in both developed and developing countries of the world, from a variety of non-agricultural uses. These inflate the value of land well above its real current agricultural value and reduce or even destroy its economic viability for agricultural production.

Pastoral farming

A very large proportion of the earth's land surface is unsuitable for cultivated crop production mainly because of a growing season limited by either insufficient rainfall or low temperatures or to a lesser extent because of steep slopes, shallow stony or inherently infertile soils. Agricultural use is hence limited to some form of pastoral farming, i.e. domestic livestock production on the basis of the natural or semi-natural ('wild') vegetation. Land used for this purpose – rangeland (in some cases termed 'rough grazing') – accounts for nearly 50 per cent of the total land area of the world today, half of which is in tropical/subtropical areas, and half in cool-temperate and cold climatic areas (Fig. 9.1). In both, however, pastoral farming is subject to comparable biophysical constraints and problems.

A large proportion of the rangeland is in the arid and semi-arid areas of the world, where the most important factor limiting primary biological productivity is water. Annual precipitation is low and seasonal with a marked summer incidence except in those areas, between the warm and cool-temperate climates, where (as in the Middle East, North Africa and Southern Australia) there is a winter rainfall maximum. The 'growing' or 'green' season, when precipitation alone or precipitation and soil moisture combined are sufficient to sustain plant growth, is short; the dry season is long and, particularly in areas of summer rainfall, drought conditions are exacerbated by high evapotranspiration rates, which reduces the amount of effective rainfall. In addition, rainfall intensity is high and as a result soil infiltration capacity rates tend to be low. More significant, however, than the shortage of water is the high degree of *rainfall variability* in both amount and incidence that is characteristic of these areas; and variability increases as rainfall amount and period of incidence decreases (see Table 9.1). Consequent upon the variability of rainfall in the semi-arid rangelands, periods of varying duration of drought when precipitation fails or is well below average are a recurrent hazard. In both temperate and tropical areas of the world the dry rangelands coincide with areas of little relief – plains, plateaus or extensive intermont basins or valleys where exposure to high wind force adds to the harshness of the climate.

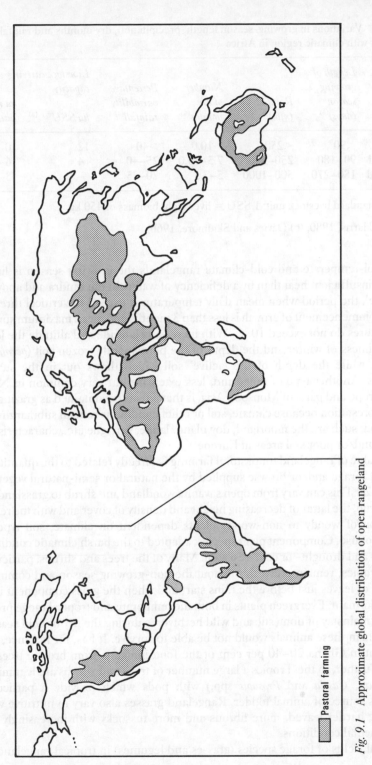

Pastoral farming Approximate global distribution of open rangeland

Fig. 9.1 Approximate global distribution of open rangeland

Table 9.1 Variations in growing-season length, precipitation, dry months and rainfall variability with climatic region in Africa

Climatic region	Length of growing season (days)[a]	Rainfall (mm)[a]	Number of dry months[a]	Percentage variability rainfall[a]	Livestock carrying capacity ha SSU^{-1} [a]	ha mature cattle^{-1} [b]
Arid	<90	<250	>10.0	>40	12	10–15
Semi-arid	90–180	250–500	7.5–10.0	25–40	4	6–10
Subhumid	180–270	500–1000	5–7.5	20–25	–	3–5

SSU = Standard livestock unit: 1 SSU = liveweight biomass of 450 kg.

(from (a) Harris, 1980, (b) Davies and Skidmore, 1966)

In cool-temperate and cold-climate rangelands the growing season is limited more by insufficient heat than by a deficiency of water. In the tundra and above the 'tree-line', the period when mean daily temperatures exceed the critical threshold for the commencement of growth is less than 3 months and mean maximum summer temperatures do not exceed 10 °C. With increasing latitude and altitude the length and harshness of winter and the depth of the permanently frozen soil (*permafrost*) increase, while the depth of the 'active' soil which thaws out in the summer decreases. Another type of rangeland, less extensive but very common in North-West Europe and parts of Monsoon Asia, is that left and maintained as grazing land after deforestation because climate, soil or relief conditions remain submarginal for cultivation; such are the moorland, downland, lande, garrigue etc., characteristic of uplands and/or poor soil areas in Europe.

The value of rangeland for pastoral farming is directly related to the quantity and quality of forage and/or browse supplied by the natural or semi-natural vegetation (Table 9.2). This can vary from open savanna woodland and shrub to grassland with dwarf shrub, the latter of decreasing height and density of cover and with the relative proportion of woody to non-woody plants dependent on climate, soil type and intensity of use. Component plant species, adapted to the harsh climatic conditions, are hardy and drought- or cold-resistant. Many of the trees and shrubs, particularly in the Tropics, remain green throughout the non-growing season and commence regrowth of leaves just before the rains start and when the grass component is still dry and dormant. Evergreen plants in both the temperate and tropical range provide the basic mainstay of domestic and wild herbivores during the dry or cold season – without them these animals would not be able to survive. It has, for instance, been noted that in Ghana 20–40 per cent of the food intake is from browse trees and shrubs. Further, in the Tropics a large number of trees and shrubs are leguminous (e.g. *Acacia*, *Cassia* and *Prosopis* spp.) with pods which provide a particularly nutritious source of animal fodder. Rangeland grasses also vary in nutritive value, becoming harder leaved, more fibrous and more tussocky with increasingly poor environmental conditions.

Annual yields of forage species (grasses and legumes) in tropical rangeland are,

Table 9.2   Annual primary productivity for major plant formations of grazing ecosystems

Latitudinal belts and plant formations	Annual productivity $(gm^{-2} yr^{-1})$
Boreal belt, humid and semi-humid regions	
Mountain meadows	750–1200
Sub-boreal belt, humid regions	
Herbaceous prairie on meadow chernozem-like soils	1500
Sub-boreal belt, semi-arid regions	
Steppe	1300–1500
Halophytic formations	400
Psammophytic formations	800
Dry steppe	900
Desert steppe	500
Mountain dry steppe	700
Mountain meadow steppe on subalpine mountain meadow steppe soils	1100
Sub-boreal belt, arid regions	
Steppified desert on brown semi-desert soils	400
Subtropical belt, humid regions	
Herbaceous prairie	1300
Subtropical belt, semi-arid regions	
Shrub-steppe	600–1000
Psammophytic formations	500
Halophytic formations	50
Mountain shrub-steppe formations on grey-brown mountain soils	800
Subtropical belt, arid regions	
Steppified desert	1000
Tropical belt, humid regions	
Seasonally humid evergreen forest and secondary tall-grass savanna on red ferralitic soils	1600
Tropical belt, semi-arid regions	
Grass and shrub-savanna soils	700–1200
Mountain savanna on red-brown mountain soils	1200
Tropical belt, arid regions	
Desert-like savanna	400

(from Caldwell, 1975)

consequent on high temperatures during the growing season, higher (35–38 tonne ha^{-1} yr^{-1}) than in temperate (20–27 tonne ha^{-1} yr^{-1}) areas. The digestibility of tropical forage, however, compares less favourably with that of the temperate range (see Table 9.3). It has been estimated that 52 per cent of all tropical grasses contain

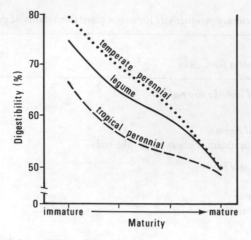

Fig. 9.2 Maturity and digestibility relationships between a representative of a temperate perennial grass species (*Dactylis glomeratus*) and a tropical perennial grass species (*Cynodon dactylon*) (from Norton, 1982)

Table 9.3 Percentage digestibility of tropical and temperate rangeland forage species

	Legumes	Grasses
Tropical	50.6	55.4
Temperate	60.7	68.2

(from Norton, 1982)

less than 9 per cent crude protein compared with 32 per cent of the temperate species. A minimum of 15 per cent is required for lactation; and less than twenty tropical species attain this level (Norton, 1982). This is reflected in differences in the mean dry-matter digestibility until flowering, while that of many tropical species (e.g. *Panicum, Chloris* and *Hyparrhenia* spp.) declines rapidly during growth (see Fig. 9.2). In addition, the leaf structure of tropical C4 grasses is characterized by a high density of vascular material which makes them more resistant to both mechanical and microbial degradation. It has also been suggested that the more erect habit of most tropical grasses (and legumes) and the lower leaf density restricts harvestability and intake by grazing animals.

Primary biological productivity of rangeland is not only relatively low but is also markedly seasonal; and seasonality of plant growth results in an annual alternation of a relatively short period of surplus forage supply with a long period of deficiency (Fig. 9.3). During the latter period livestock may, at best, have just sufficient to maintain themselves. More frequently, available feed is insufficient for maintenance requirements, and animals lose condition and weight. In addition, the available forage, particularly that supplied by grass, declines rapidly in nutritive value as the plants mature. In the case of Scottish hill-farms it has been shown that the primary

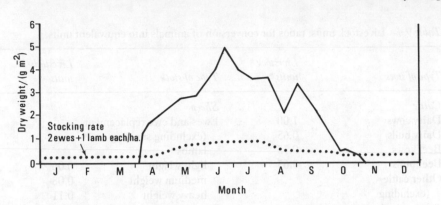

Fig. 9.3 Daily grass requirements of sheep compared with daily growth rates of a grass on a hill site in the UK (from Morris, 1977)

nutritional constraint on animal productivity (an output of weaned lambs of 15.3 kg $ha^{-1} yr^{-1}$ compared to 612 kg $ha^{-1} yr^{-1}$ from the lowlands) is an inadequate winter intake of feed (Eadie and Cunningham, 1971).

The type of livestock, the use to which it is put and its productivity in terms of human food production are dependent on the severity of the environmental conditions on the one hand and the particular pastoral system on the other. The efficiency with which domestic livestock can use available forage and maintain themselves varies. Goats, sheep and particularly camels, llama, yak and reindeer, adapted to harsh environments, are better able to digest and maintain themselves on a poor diet than cattle. The number of livestock units or livestock equivalents (see Table 9.4) that can, ideally, be maintained without loss in productivity (i.e. that will ensure sustained production) on a given area is referred to as its *carrying capacity*.

Carrying capacity, however, is difficult to express in quantitative terms. It is often more an expression of individual or collective knowledge of the relationship between livestock numbers and range condition. *Stocking rate*, on the other hand, is a measure of actual use, i.e. the number of animals on a particular range or grazing area of a given extent for all or part of the year. The stocking rate of rangeland is characteristically low; and as primary biological productivity declines the area needed to supply forage for one animal increases. *Stocking density* is a measure of livestock 'crowding' and is usually calculated on a daily basis. Stocking density is closely related to *grazing density* or *pressure* and the availability of drinking water. To a very considerable extent the need for water determines the grazing range of different types of livestock, particularly in arid or semi-arid environments.

Water is vital for the survival of cattle and the amount needed varies from 5 to 23 litres day^{-1} per head mounting with increased growth rate, temperature, lactation and movement; lactating livestock need up to 44 per cent more water, depending on breed. Lack of water reduces appetite and depresses voluntary food intake. Experiments in Australia have shown that the density of grazing decreases in proportion to the square of the distance from the water source (Jeans, 1977). Need for water limits the daily grazing range to *c.* 13 km though watering on alternate days may, of necessity, be imposed on nomadic or transhumant animals (see Table 9.5). Cattle can graze up to 40 km, while camels have a range of 80 km. There are

Table 9.4   Livestock units; ratios for conversion of animals into equivalent units

Type of stock	Livestock units*	Type of stock	Livestock units
Cattle		*Sheep*	
Dairy cows	1.00	Ewes and ewe replacements	
Dairy bulls	0.65	(excluding suckling	
Beef cows	0.75	lambs	
Beef bulls	0.65	light weight	0.06
Other cattle		medium weight	0.08
(excluding		heavy weight	0.11
intensive beef systems)		Rams	0.08
0–12 months	0.34	Lambs	
12–24 months	0.65	birth to store	0.04
>24 months†	0.80	birth to fat	0.04
Barley-beef	0.47	birth to hoggets	0.08
Horses	0.80	purchased stores	0.04

* A *livestock unit* is usually defined in terms of feed requirements. The ratios in the table are based on metabolizable energy requirements, with one unit being considered as the maintenance of a mature 625 kg Friesian cow and the production of a 40–45 kg calf, and 4500 litres of milk at $36\,g\,kg^{-1}$ of butterfat and $86\,g\,kg^{-1}$ solids non-fat.
† Reduced in proportion to time animal is on farm.

(from Scottish Agricultural Colleges, 1987)

rangeland areas where a lack of water sources or water with an unpalatably high mineral level can limit use for grazing entirely during a greater part of the year.

Animals adapted to arid environments have a lower *water turnover* (quantity or percentage water used in a given time unit); according to Nicholson (1985) the range in turnover rates is comparable in zebu cattle, sheep and goats, i.e. *c.* $70–200\,ml$ $kg^{-1}\,day^{-1}$; in contrast, that of the camel is $37–70\,ml\,kg^{-1}\,day^{-1}$. The latter void faeces with 45 per cent dry matter; cattle, with 30 per cent. The bedouin goat appears to be most highly adapted to large water intake, imbibing on average 30 per cent of its hydrated weight. While the camel has a low water turnover rate, size coupled with infrequent visits to water-holes means that its total intake at any one time is very large – up to 100 litres. In addition, the camel's tolerance of saline drinking water is higher than that of any other grazing animal (5.5 units of total dissolved salts are tolerated by the camel compared with 1.5 units by the goat, 1.3–2.0 units by the sheep and 1.0–1.5 units by cattle).

Competition for forage from wild herbivores can be a significant though variable factor in determining the carrying capacity of rangeland for domestic livestock. Large wild grazing and browsing herbivores are most numerous in the African savannas and in Australia, where the marsupials occupy a similar ecological niche to the mammals in Africa. However, grazing and burrowing rodents – such as rabbits and hares – together with feral herbivores constitute serious competitors with domestic livestock in the temperate rangelands. This is particularly the case where

Table 9.5 Drinking-water requirements for selected grazing livestock assuming sufficient feed

Livestock	Maximum summer period without water (days)	Radius range from waterpoint*
Sheep/goats	3–5	1.0
Donkeys	4	1.3
Camels	12	4.0
Cattle	0.4	2.0

* Radius range is given as a ratio to that for sheep; thus the radius range for sheep is 1.0 and that for the camel, whose range is four times that of the sheep, 4.0.

(from Schmit-Neilson, 1956)

natural predators have been eliminated or, as in New Zealand, did not exist when livestock were introduced from Europe. Research in Australia has revealed the degree of competition for herbage from termite (*Drepainotermus* spp.) grazing, with up to 375 kg ha^{-1} of grass lost in humid years (Joss *et al*, 1987). Competition has the greatest impact on carrying capacity where there is a high degree of overlap in habitat and species preference between wild and domesticated animals (Fig. 9.4).

The extent to which management can rectify the inherent imbalance between seasonal forage production and constant annual demand is limited. Grazed herbage – particularly grassland – is more productive than that not grazed. Defoliation stimulates growth. However, production is usually in excess of that needed during the growing season and the accumulation and slow decay of uneaten herbage can depress subsequent production. Consequently burning as a result of either 'wild' or 'set' fires is an important environmental factor in both tropical and temperate rangeland. The frequency of fires started by lightning varies with the amount of combustible litter and the amount and intensity of precipitation accompanying storms. In Australia it has been estimated that *c.* 6 per cent of all bush fires are started by lightning. Prescribed or controlled burning of rangeland is particularly characteristic of the more sparsely populated savannas of Australia and South America – where annual productivity may be high but the nutrient content of the grasses is low (Harris, 1980). Fire is, nevertheless, also frequently used in the more densely populated African savannas as well as in the more arid desert grasslands of North America and Eurasia.

Burning clears off dry useless herbage and by releasing nutrients stimulates the productivity of fresh nutritious growth when climatic conditions allow. It is also an effective weed, pest and disease controller. The use of fire, however, gives a competitive advantage to fire-tolerant plant species, more particularly to herbaceous (especially grasses) at the expense of woody forms. On the other hand, a decrease in burning either because of deliberate fire control or because of a decline in grass yield consequent on grazing pressures can, in arid rangelands, result in an increase in fire-intolerant desert scrub species (e.g. mesquite, *Propopis juliflora*; creosote, *Larrea tridentia*; acacia, *Acacia* spp.; scrub-pine, *Pinus ponderosa*) at the expense of grasses.

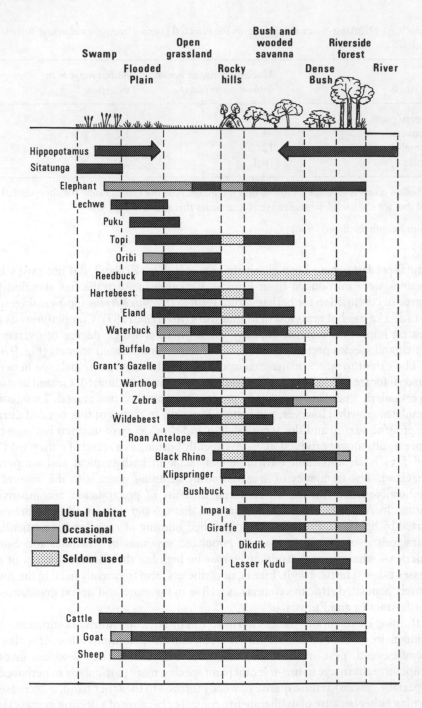

Swamp Open grassland Bush and wooded savanna Riverside forest

Flooded Plain Rocky hills Dense Bush River

Hippopotamus
Sitatunga
Elephant
Lechwe
Puku
Topi
Oribi
Reedbuck
Hartebeest
Eland
Waterbuck
Buffalo
Grant's Gazelle
Warthog
Zebra
Wildebeest
Roan Antelope
Black Rhino
Klipspringer
Bushbuck
Impala
Giraffe
Dikdik
Lesser Kudu

Cattle
Goat
Sheep

■ Usual habitat
▨ Occasional excursions
▦ Seldom used

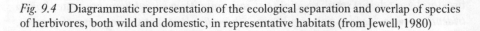

Fig. 9.4 Diagrammatic representation of the ecological separation and overlap of species of herbivores, both wild and domestic, in representative habitats (from Jewell, 1980)

The burning rotation on cool-temperate northern rangeland, dominated by slow-growing woody heath species, is usually longer, varying from 7 to 15 years. In this case the aim is to burn the vegetation near the end of its 'building' stage and hence maintain as high a level of productivity of green shoots as possible. The proportion of rangeland vegetation burned in any one year will be determined by the length of the burning rotation. In addition, fire can be used at different seasons with differing effects on the vegetation. Early dry-season burns are usually light; late burns are heavier; while those during the period of early rains will affect scrub and tree growth rather than the still-dormant grasses, hence giving the latter a competitive advantage.

Table 9.6 Relationship between type of livestock and rainfall in Africa

Rainfall (mm yr⁻¹)	Predominant grazing system	Predominant animal
<50	occasional nomadic	camel
50–200	nomadism with long migration	camel
200–400	all types nomadism; transhumance plus supplementary arable farming	cattle goats sheep
400–600	semi-nomadism, transhumance; partial nomadism; emphasis on arable farming	cattle goats sheep
600–1000	transhumance and partial nomadism	cattle
>1000	partial nomadism and permanent stock-keeping	cattle

(from Ruthenberg, 1976)

One of the most important adaptations of the grazing livestock to seasonal and spatial variation in forage is by migration from areas of low to those of high or higher productivity (see Table 9.6). *Nomadism*, in which herding is adjusted to the variations in the seasonal occurrence of rainfall and vegetation, can be total or partial. The former is restricted to those areas marginal to the desert where the carrying capacity and water availability are insufficient to maintain a sedentary way of life. The camel, because of its high degree of adaptation to drought, is the most important component of nomadic herds, with sheep, goats and, to a lesser extent, cattle in less rigorous conditions. Nomadic herding occurs mainly in Africa, and the Middle and Far East, never having evolved, or having been displaced, in the New World rangelands. However, nomadism is declining and partial or semi-nomadism is now more common than total nomadism, particularly in the African savannas and among the remnants of the Navajo sheep-herders in Arizona. In these cases the herds and herdsmen migrate between the rangeland and permanent settlements. Movement may be seasonal, annual or even biannual associated with temporary kraals built beside the more reliable water supplies; spatially it may be horizontal or vertical (see Fig. 9.5). The former is more characteristic of the African pastoral systems; the latter of mountain areas in Europe and North America. Regular seasonal vertical movement of grazing animals, related to the availability of different types of rangeland, has long been a feature of pastoral farming in upland and mountain areas in Europe and, more recently, of livestock ranching in North America. The movement of domestic livestock and people to 'summer pastures', at a shorter or greater distance from the 'home farm', is associated with variation in length of the growing or grazing season with altitude and is called *transhumance.* The montane grasslands above the tree-line of the Norwegian saeters and the Swiss alps are cases in point. In Sweden the mountain herds of reindeer migrate or are transported longer distances between their summer pastures on the high fjells and winter forest-grazings by the coast, by way of established autumn and (on their return) spring grazings, where calving takes place. Transhumance on a more extensive scale is still characteristic of the basin and range area of the North American Rockies. The seasonal grazing varies from place to place. The three main types of rangeland, dependent on altitude, are:

1. *Summer range* in high mountains; this provides a short 3-monthly (July to September) grazing season between snow-melt and snowfall with a potentially high carrying capacity which provides animals with a nutritional 'boost' before the food-deficient winter season.
2. *Spring/fall ranges* at lower altitudes have higher temperatures but often less precipitation than the high mountain pastures. These can be used earlier in spring and later in autumn, either *en route* to or from the summer range or continuously for the whole of a 6-month growing season from mid-April to mid-October.
3. *Winter range* is that associated with the lowlands and lower intermont valley and basin floors where mean annual precipitation is even lower and diurnal and, in some cases, annual ranges of temperature are great. Provided water is available they can be grazed all year; however, productivity is limited by drought throughout the area and by extreme winter cold in the northern ranges.

The amount of different kinds of vegetation within a grazing area is as important a determinant of carrying capacity and livestock productivity as is the total amount of range forage. This is reflected in the hill sheep farming of Scotland, albeit on a smaller scale. The Scottish hill areas with a variety of plant communities which can provide perennial evergreen winter forage and an early spring 'bite', a well as the more nutritious summer grassland, are more useful and productive than rough grazing dominated by one type of forage alone.

LIVESTOCK PRODUCTIVITY

On most of the rangelands, grazing of domestic livestock is year-long and uncontrolled, i.e. it is 'open' or 'free range', with the numbers of animals and their productivity in terms of meat, milk, hides, wool or fibre closely related to the climatically determined forage and water resources. The type of animal and of

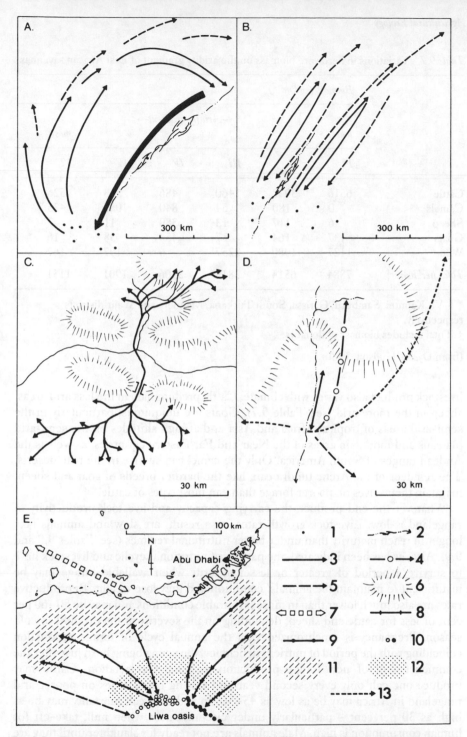

Fig. 9.5 Types of nomadic movement: A) *horizontal* (elliptical); B) *horizontal* (pulsatory); C) and D) *vertical*; E) *Bedouin routes* in Abu Dhabi. 1. Oases; 2. Wadis; 3. Normal routes; 4. Good year routes, abundant forage; 5. Bad year routes, scarce forage; 6. Mountains; 7. Pearl fishing boat movements; 8. Tribal ('Bani Yas') livestock movements; 9. Tribal ('Nanasir') livestock movements; 10. Date gardens; 11. Winter grazing areas; 12. Summer grazing areas; 13. Bad year movements (from Heathcote, 1983)

Table 9.7 Variations in herbivore biomass on the aridity gradient of East African savannas

	Biomass ($kg\,km^{-2}$)					
			←Aridity gradient→			
	least					most
	I*	II	III	IV	V	VI
Cattle	6516	3942	1460	486	553	526
Camels	0	180	64	840	1052	435
Sheep	6	117	13	380	11	7
Goats	137	164	17	454	38	16
Wild	1127	1980	2257	6	197	167
Total herbivores†	7884	6514	3818	2406	1901	1151

* I–VI: Kaputei, Samburu, Garissa, South Turkana, Mandera and Wajir districts respectively.
† Total includes biomass of donkeys.

(from Dyson-Hudson, 1980)

livestock product also varies with climate. Cattle predominate in the less arid areas, sheep in the more arid (see Table 9.7). Goats are ubiquitous, particularly in the semi-arid areas of tropical Africa and Asia and at high altitudes on the semi-arid plateaus and mountain areas of the Near and Far Eastern countries, as well as the Andean ranges of South America. Only the camel can survive in the true deserts. The reindeer of the Arctic tundra can, like the hardier breeds of goat and sheep, maintain themselves on poorer forage than can most types of cattle.

Whatever the end-product, the carrying capacity and livestock productivity of rangeland is low. Livestock growth rates, as a result, are slow and animals take longer to reach maturity than under better nutritional regimes (see Tables 9.8 and 9.9). Also, as has been indicated, the pattern of nutrition is cyclic and livestock have to survive a period of greater or lesser severity when available forage may be insufficient to maintain the animals' condition and they lose weight. Reproductive rates are also much lower than in more favourable conditions and may be 50–60 per cent or less for cattle and sheep, depending on the severity of the drought or cold season. Pregnancy is synchronized with the annual cycle of forage production coinciding with the period of nutrient deficiency, the young being born just after the commencement of new forage production essential for lactation. Cattle may produce one calf only every second year and calving percentages on poorer arid rangeland in Africa may be as low as 45–50. In addition, calf mortality may be as high as 30 per cent – particularly under circumstances when milk take-off for human consumption is high. Male animals are not ready for slaughter until they are 4–5 years old. On African rangeland, productivity is low in terms of the percentage extraction rate, i.e. cattle slaughtered as a percentage of the total herd (6.2–21.5 per cent), carcass weight (90–120 kg) and off-take by carcass per kilogram per head of

Table 9.8 Production parameters for sheep and goats on arid rangeland in Africa

	Mali		Kenya	
	Sheep	Goats	Sheep	Goats
Average age at first conception (days)	474	403	527	576
Weight at first conception (kg)	23.4	16.9	–	–
Breeding interval (days)	250	270	320	320
Litter size	1.05	1.21	1.04	1.25
Number young per year	1.51	1.62	1.19	1.42
Growth rate at 150 days (g day^{-1})	89	61	74	57
Preweaning mortality	31	35	33	32

(from Butterworth and Lambourne, 1987)

Table 9.9 Production parameters for cattle on South American rangeland

	Brazil	Colombia	Venezuela
Average stocking rate (animal units ha^{-1})	0.23	0.17	0.32
Heifer weight at 36 months (kg)	283	255	270
Age at first conception (months)	40	35	38
Weaning rate (%)	57	45	52
Liveweight gain (animal units yr^{-1})	65	58	56
Liveweight gain (ha yr^{-1})	12	12	32

(from Vera et al., 1987)

the total herd (14.7 ± 6.1). Further, temperatures above or below the limits of comfort can have debilitating effects on livestock which, combined with a starvation or near-starvation diet, make them particularly susceptible to disease.

The impact of disease on livestock distribution and productivity is most severe in the intertropical regions of the world. In addition to the diseases introduced from temperate areas, there is a large variety of either free-living pathogenic organisms or of insect- and tick-borne diseases. Many are fatal and can destroy whole herds. Others are difficult to combat since they involve a two- or three-host life-cycle outside the domestic animal. Further, in some instances, feral animals such as water buffalo, camels, donkeys, goats and pigs as well as game animals provide disease reservoirs from which domestic livestock may be infected. Among the most virulent tropical animal diseases are rinderpest and, more particularly, animal trypanosomiasis. It is estimated that approximately twelve million square metres of African savanna, between 14°S and 29°S and from sea level to 1800 m OD, are infested with tsetse fly (Glossina spp.) of which there are twenty-two recorded species each restricted to a particular habitat. In about two-thirds of East Africa, half of West Africa and a quarter of Central Africa tsetse-borne disease limits and usually prohibits cattle rearing (Jordan, 1980). Even if cattle and humans are removed from infected areas, a reservoir of infection is maintained by the wild game.

The type of livestock and the productivity of the rangeland depend on the particular pastoral system. Temperate and tropical rangelands in the agriculturally developed countries are dominated by large individually owned breeding herds of mainly cattle and/or sheep. The animals that form the domestic livestock of today were not indigenous to the Americas and Australasia. Exotic strains of mainly European sheep and cattle (with breeds of *Bos indicus* in Australia) were introduced mainly in the nineteenth century; and from these, new strains, better adapted to local environmental conditions, have since been developed. As Heathcote (1983) notes, the temperate system of livestock management developed in Europe was, only slightly modified, imported into the arid lands. The aim of this commercial ranching was and still is to produce meat 'on the hoof' – store or immature animals – which are sold off the rangeland for finishing or cross-breeding elsewhere. As demand for meat in the Western world grew, so the ranchers pushed further into the semi-arid and arid rangelands.

Initially the most sought-after land was that with access to surface water though, later, surface storage and, more importantly, the use of deep-drilling techniques to tap ground water helped to mitigate the problem of seasonal and long-term drought. In some areas there was sufficient water for irrigation and the production of supplementary feed. In the recurring periods of disastrous weather, stock had either to be moved to alternative (often leased) rangeland in the mountains or sold off in order to reduce the stocking rates to a level commensurate with the lowered carrying capacity of the range.

Overstocking and deterioration of range conditions (often exacerbated as in Australia by introduced pests such as the rabbit) and soil erosion ensued. In the USA it was estimated that by 1930 the carrying capacity of the western range had fallen by 50 per cent. Legislation was introduced to regulate grazing on publicly owned land; and, since the 1940s, research has concentrated on problems of alleviating seasonal variations in natural forage production, ameliorating the effect of drought or cold hazards, controlling disease and preventing soil erosion in order to maintain the range in as good a condition as possible. There has been the application of capital and modern technology to the development of ground-water resources, methods of grazing management and methods of pasture improvement, combined with the movement of herds by road or rail, the importation of supplementary feed and/or the feeding of young stock elsewhere during periods of annual forage deficiency. Despite these developments ranching is still dependent on the natural resource base and both primary and secondary production remain low. Indeed it has been noted by (Joss *et al*, 1987) that the annual cattle 'turn-off' (i.e., number of animals killed or taken off the total herd) of 9–19 per cent in Central Australia in the 1970s was little different from that of the Masai herds in Africa.

In the subsistence pastoral systems associated with the tropical rangelands, particularly of the Third World countries, the most important product is milk. A greater variety of livestock is kept than on the temperate rangelands and all are, to a lesser or greater degree, multiple-purpose animals, the same animal being used for draught, milk and meat, and animal dung providing a source of domestic fuel as well as fertilizer. Where there is a high dependence on milk, the number of animals needed to support a family or social group is large, partly because of the necessity to insure against poor years when mortality can exceed 70 per cent and partly because production is relatively inefficient: the cow takes a long time to mature and calving rates are low. In addition, there is a retention of older animals in excess of those required for breeding or work, while the ratio of male to female animals is high. This makes for a different herd composition than on temperate rangeland. Further, domestic livestock, particularly cattle, have a considerable social or status value which is dependent on their numbers rather than on their quality or productivity. This results in higher stocking rates than on temperate ranges. Also, in contrast to commercial ranching, while livestock are owned by individuals and/or families, rights to water and grazing are not. As a result it has been suggested by Tribe *et al* (1970) that land use tends to be opportunistic and to be determined by immediate environmental conditions and short-term objectives. Nevertheless, Brachett (1987) argues that communal pastoral systems based on rangeland in Africa produce more food per unit area than any other animal-based arid land-use system: e.g. 1.7 and 3.2 kg protein $ha^{-1} yr^{-1}$ in the Nigeria Delta and Southern Ethiopia compared with 0.4–0.5 kg from Australian ranches in similar rainfall zones.

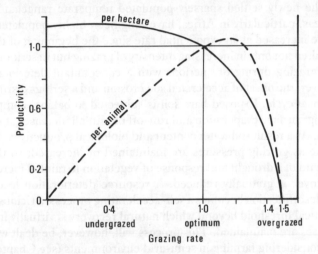

Fig. 9.6 Optimal grazing rate (from Heady, 1975)

Common to all rangeland used for pastoral farming is the risk of resource deterioration as a result of *overgrazing* (see Fig. 9.6). This involves a decline in productivity of the vegetation in terms of both quality and quantity. All grazing and browsing livestock are, to a lesser or greater degree, selective feeders. Selectivity may be a function of eating method: some such as cattle pull (i.e. tear up) the forage, hence preferring and being better adapted to cope with tall-grass range; others such as sheep and rabbits nibble and hence favour a shorter sward. Cattle have a more marked preference for grasses, while sheep and goats have a slightly higher tendency to browse given the opportunity. Mixed stocking makes for maximum use of browse, field-layer shrubs and grasses throughout the year and modifies the effects of high selectivity by the type of animal. Selectivity is also a function of the relative palatability of the constituent forage species, the more nutritious and more palatable being eaten in preference to the less nutritious and less palatable. As a result the latter tend, over time, to gain in abundance at the expense of the former. With increased intensity of grazing, defoliation exceeds replacement by growth and a quantitative decline in forage is followed by progressive weakening and removal of the vegetation cover leading to increasing susceptibility to accelerated soil erosion.

This process of resource deterioration is also exacerbated by overburning and drought. Overburning is a result of too great an intensity of fire, usually as a result of the accumulation of combustible fuel in vegetation that has not been burned for some time, combined with low humidity and high wind force. If the soil organic matter as well as underground storage organs are destroyed, recovery is slow and precarious because of grazing pressures and/or the acceleration of soil erosion in areas stripped bare of a protective organic cover. In the semi-arid and arid range-lands the main hazard is drought. Protracted periods – a run of years – when rainfall is below average or nil is characteristic. During droughts the quantity and quality of forage and hence its livestock-carrying capacity declines. Livestock numbers may be reduced naturally by starvation, by out-migration or by deliberate reduction. In contrast to the newly settled sparsely populated temperate rangelands, those of semi-arid areas, particularly in Africa, have long been highly populated and stock numbers have increased at an exponential rate since the beginning of this century. This has resulted not only in increasing intensity of grazing but in serious overstock-ing and overgrazing during dry periods with a concomitant deterioration in the vegetation cover, the onset of accelerated soil erosion and a serious disruption of the soil hydrological cycle. Exposed bare soil is subjected to baking, compaction and surface sealing; surface evaporation and run-off of rainfall increases at the expense of infiltration. As a result soil water content and biological productivity decreases at the same time as grazing pressures are maintained or increased; so the time-lag between the end of a drought and response of vegetation to rainfall increases, while ability to recover is gradually reduced. A resource deterioration leading to soil erosion and desertification becomes a self-accelerating process, a vicious downward spiral to a critical threshold beyond which natural recovery is virtually impossible if existing stresses are maintained. This process will, however, be dealt with in more detail when considering farming in semi-arid environments (see Chapter 12).

10

The humid tropical lowlands

The Tropics, as the name implies, is that circumequatorial area between latitude 23½°N and 23½°S. It comprises some 40 per cent of the total land area and contains about the same proportion of the world's population. It is an area of potentially high biological productivity but low agricultural yields relative to other areas of the world. The interaction of climate and soil has produced a resource base, less easy to manage and conserve than elsewhere and which creates agricultural problems differing in nature and intensity from those in more temperate climatic regions of the world.

Throughout the humid tropical lowlands insolation and air temperatures are constantly high. There is little variation in day length throughout the year. Diurnal and annual ranges of temperature are small – the former usually less than the latter. The Tropics is the 'hot' part of the earth and Nienwolt (1975) uses two parameters to define the boundaries of this climatic zone: a mean minimum monthly temperature of not less than 18 °C and a comparable mean annual and mean monthly range of temperature (Fig. 10.1). The respective isolines show a general spatial coincidence – greater in the northern than in the southern hemisphere. Within the Tropics so defined, frost is unknown and temperatures are rarely low enough to inhibit plant growth. They can, however, preclude the successful cultivation of temperate crops which require lower optima, often combined with much greater seasonal and diurnal variations in temperature and day-length than occur in the Tropics.

Extensive cloud-cover reduces the amount of insolation (as well as of outgoing radiation) to levels lower than in higher latitudes where, in addition, day-length during the growing season is much longer; indeed the amount of insolation received during the growing season in temperate latitudes is higher than in the Tropics. Nevertheless, where there is no deficiency of water, potential photosynthesis and consequently plant productivity are two to three times higher in the humid Tropics because of a year-long thermal growing season. Deficiency of soil water is the main environmental factor limiting crop and livestock production in the Tropics. Over

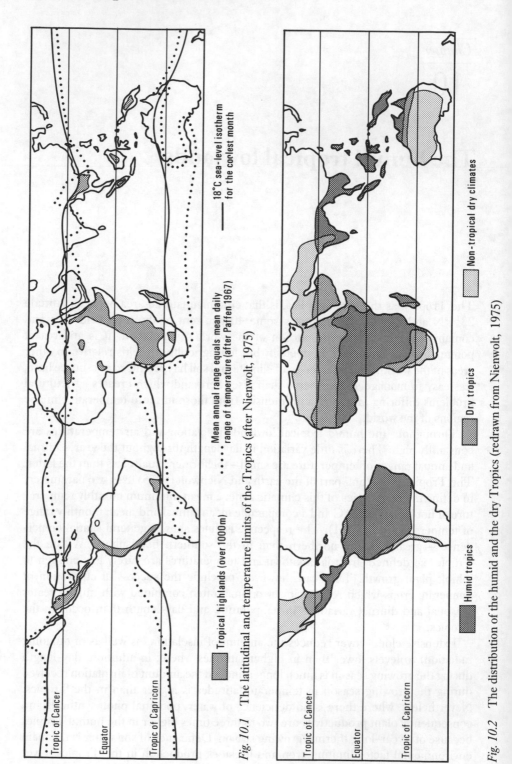

Tropical highlands (over 1000 m)

······· Mean annual range equals mean daily range of temperature (after Paffen 1967)

—— 18°C sea-level isotherm for the coolest month

Fig. 10.1 The latitudinal and temperature limits of the Tropics (after Nienwolt, 1975)

Tropic of Cancer

Equator

Tropic of Capricorn

Humid tropics

Dry tropics

Non-tropical dry climates

Fig. 10.2 The distribution of the humid and the dry Tropics (redrawn from Nienwolt, 1975)

much of the area, distribution of precipitation (Figs 10.2–10.4) is markedly seasonal, with a clear alternation of one or two humid periods when rainfall exceeds potential evapotranspiration with one or two dry periods when rainfall is less than potential evapotranspiration and a soil water deficit results (Fig. 10.5). Constantly high evaporation rates further reduce the amount of effective rainfall reaching the plant root zone; and, as many authors note, an amount of precipitation that would be more than sufficient for agriculture in temperate regions may be quite inadequate in the Tropics. The amount of rainfall and the relative length of the dry and wet seasons are the major determinants of agricultural potential in terms of products and yields.

The rainfall regime (see Table 10.1) has been the basis for most attempts to classify tropical climates from an agronomic standpoint, with the main distinction usually being made between humid regimes with a very short or no really marked dry season; semi-humid regimes with a marked wet and dry season of varying length; and dry regimes, where the rainy season is very short and very variable in length and in amount of precipitation. The semihumid wet and dry climate is the most extensively represented of the three regimes in the Tropics. Within it, however, there is a continuous variation from wetter to drier areas with distance from the influence of either equatorial or tropical monsoonal air masses as the case may be.

Marked climatic seasonality is accompanied by a high degree of variability (particularly of the amount of annual rainfall, the beginning of the rainy season and the duration of the humid growing season) which assumes an even greater agricultural significance than in temperate regions. A high degree of seasonality is further exacerbated by the nature and intensity of tropical rainfall (Fig. 10.6). A high percentage of the annual precipitation is concentrated in violent storms or heavy rainfall events of relatively short duration. Intensity of rainfall is correspondingly high with rates of fall of 150 mm h^{-1} commonly occurring. In comparison, rainfall intensity only exceeds half this rate on rare occasions in summer thunderstorms in temperate regions. High intensity reduces the effectiveness of tropical precipitation since it tends to exceed the infiltration capacity of the soil. As a result loss by surface run-off, particularly on bare ground, increases flash-flooding on often extensive areas of low gradient; and impeded soil drainage occurs. Erosivity of rainfall increases above an intensity threshold of 25 mm h^{-1}, which is equivalent to the physical impact of $c.\ 250 \text{ tons h}^{-1}$ of water. Soil erosion risks are consequently much greater than in humid temperate areas. Tropical farmers are bedevilled by the twin problems of water surpluses and water deficits – of flood and drought – in the course of one year. These can be further intensified by high wind forces which accompany tropical cyclones and tornadoes in the wet season and dust-laden desiccating winds in the dry season.

PRODUCTIVITY

Despite the combination of high temperature and humidity during the growing season, crop yields in the humid Tropics are generally low and very variable in comparison to those of similar types of crops in non-tropical areas (Table 10.2).

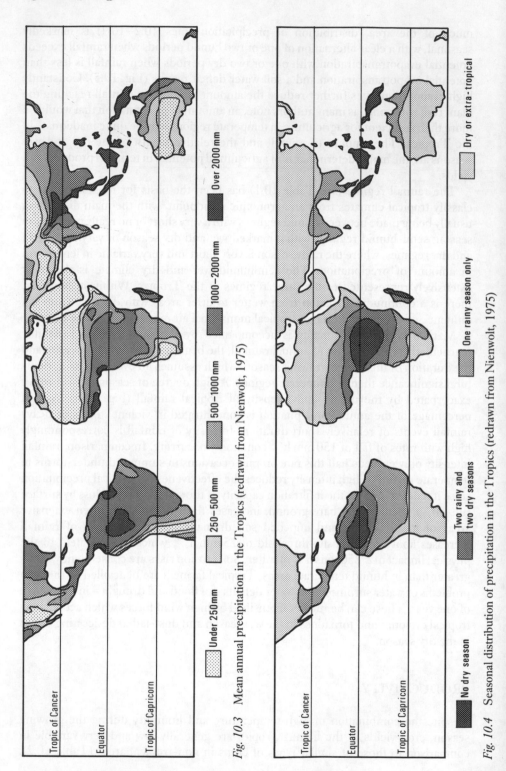

Fig. 10.3 Mean annual precipitation in the Tropics (redrawn from Nienwolt, 1975)

Over 2000 mm

1000–2000 mm

500–1000 mm

250–500 mm

Under 250 mm

Fig. 10.4 Seasonal distribution of precipitation in the Tropics (redrawn from Nienwolt, 1975)

Dry or extra-tropical

One rainy season only

Two rainy and two dry seasons

No dry season

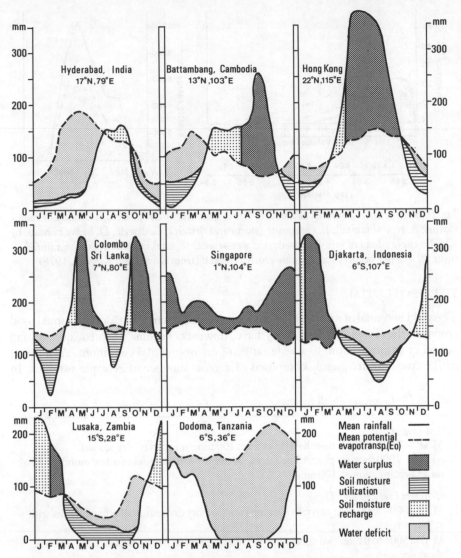

Fig. 10.5 Tropical soil water regimes (redrawn from Nienwolt, 1975)

Agriculture is characterized by low-intensity and near-subsistence farming systems. Input levels – be it of water, fertilizers, herbicides, pesticides or improved HYVs – are correspondingly very low. Crop yield, consequently, is more dependent on natural environmental conditions than in more intensively farmed areas of the world. The four major determinants of yield are:

1. The yield potential of the cultivars used.
2. The level of weed competition.
3. The intensity of pest and disease infestation.
4. The productive potential of the soil.

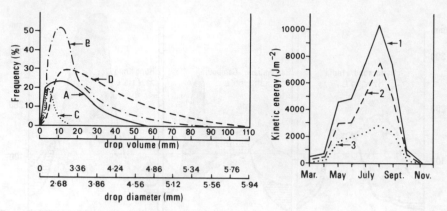

Fig. 10.6 Intensity of rainfall at Samaru, North Nigeria. a) Drop size distribution of rain-storms; A, typical rainfall; B, composite (storm and drizzle); C, drizzle; D, heavy rainfall. b) Kinetic energy-load of rain-storms during wet-season: 1, total rainfall; 2, erosive rainfall (over 20 mm per occurrence); 3, non-erosive rainfall (from Kowal and Kassam, 1978)

YIELD POTENTIAL

The yield potential of most indigenous native species, particularly of the annual food crops, in the Tropics is low. They have, however, become adapted, over a long period of domestication, to a wide range of environmental conditions. As a result native crops are frequently composed of a great number of ecotypic varieties. In

Table 10.1 Tropical rainfall regimes

Humid Tropics
1. MAP > 2000 mm; all months with at least 100 mm; no marked dry season.
2. MAP < 2000 mm; no pronounced dry season but with a period of a few months when mean precipitation < 100 mm.

Semi-humid (wet and dry) Tropics
1. MAP 2000–1500 mm; two rainy seasons and a short dry season of a few months with < 50 mm.
2. MAP 1500–650 mm; two short rainy seasons separated by one pronounced and one short dry season.
3. MAP 1500–650 mm; one fairly long rainy season *c.* 3–5 months, each with >75 mm, and one long dry season.
4. MAP 650–250 mm; one short rainy season of 3–4 months, each with >50 mm, and one long dry season.

Tropical monsoon
MAP >1500 mm; one season very heavy rain and one long dry season.

Semi-arid or dry Tropics
MAP <250 mm concentrated in 2–4 months.

MAP = mean annual precipitation.

(from Jackson, 1977)

168

Table 10.2   Average yield of tropical cereals 1978–80

	Yield (kg ha^{-1})		Tropical yield as percentage of world yield
	Tropics	World	
Rice (paddy)	2014	2723	74.0
Maize	1238	3124	39.6
Sorghum	832	1346	61.8
Millet	546	667	81.8
Others	1348	1873	72.0
Total	1403	2152	65.0

(from Norman *et al.*, 1984)

addition, many of these local crop varieties are genetically extremely diverse. In any plant population there may be differences in growth habit; in response to climate; in resistance to disease; and in yield. Hence the disadvantages of low yield potential are to a certain extent compensated by the inherent insurance against an environmental hazard such as drought or disease better provided by a genetically diverse crop than by a genetically more uniform one.

WEEDS

In addition to the agricultural problems associated with the intensity of tropical rainfall are those of rampant weed growth. Plant diversity is high in the hot and particularly humid Tropics. The pool of potential weed species is large. Competition with cultivated crops is severe and the consequent reduction of yield is much greater than in other parts of the world. Many tropical weeds are adapted to survive drought and the annual burning of fallow or grazing land. The weeds rated the nine worst in the world all occur in the Tropics. They include:

Cyperus rotundus – nut grass
Cynodon dactylon – Bermuda grass
Echinochloa crus-galli – barnyard grass
Eleusine indica – goose grass or fowl-foot grass
Sorghum halepense – Johnson grass
Imperata cylindrica – lalang
Eichornia crassipes – water hyacinth
Portulaca oleracea – purslane

Six of these are perennial grasses whose growth from underground stems or rhizomes makes them difficult to eradicate, while their vigour and spread are often stimulated by cutting and burning. Parasitic weeds also pose a serious problem in the Tropics, particularly on soils of low fertility where the host crops have been grown too frequently. These weeds are particularly persistent because their seeds have a long dormancy and can remain in the soil until the host plant is sown. The most serious of the parasitic weeds are broom-rape (*Orobanche* spp.), in drier areas on broad-leaved crops such as tobacco, beans, cotton and certain vegetables, and *Striga* spp., found particularly in Africa and Asia, which mainly infests grasses and the closely related cereals (maize, sorghum, millet). Infestations are often very heavy; Wrigley (1981) records two million *Striga* plants per hectare equivalent to a weight of over 3 tonne. It is not therefore surprising that half the input of labour for crop production in the Tropics is devoted to weeding. Other mechanical and chemical means of weeding are still too costly for the majority of tropical cultivators. In any case mechanized weeding can increase the liability to soil erosion, while the use of herbicides carries the risk, as has happened in temperate areas, of eliminating more susceptible broad-leaved weeds and thereby encouraging the spread of the previously rarer grass weeds, which are not only more competitive but are more difficult and costly to control.

PESTS AND DISEASES

Pests and diseases of both crops and livestock are also more numerous and have a more devastating impact on agricultural production in the Tropics than elsewhere. While there may be, in general, between five and twenty major insect pests which attack any crop, Hill and Waller (1985) note that some of the important cash crops of the Tropics (e.g. cocoa, cotton, coffee) may be prone to damage by as many as 500–700. It has been estimated that approximately 15 per cent of the total world crop is lost as a result of pests (with an additional loss of 20 per cent during postharvest storage). The Tropics experiences a combined crop loss of 50 per cent (20 per cent during cultivation and 30 per cent during storage). Climatic conditions are particularly favourable for the rapid proliferation of pests during the wet season. Insects and pathogenic organisms with very short life-cycles can produce several generations in one season, while the intervening dry season provides optimum conditions for the survival of spores, larvae etc. Indeed in the humid Tropics most pests can breed continuously with all an organism's developmental stages being present most of the time.

Unfortunately, the eggs and pupae of many insect pests are very resistant to contact-insecticide sprays, while the translocation and degradation of insecticides are more rapid in the humid Tropics than elsewhere. Hence chemical treatment is much less effective than in other climatic regimes. Hill and Waller (1985) also point out that short-term annual tropical crops are more subject to serious pest outbreaks ('explosions') just because periods of great suitability alternate with those that are unsuitable for the propagation and growth of certain pest populations.

Monocultures of cash crops are particularly liable to serious reduction or near destruction by pests and disease. Recent examples are the decimation of the arabica coffee industry in Sri Lanka and of the cocoa palm in Nigeria. Livestock are similarly at risk, and a large battery of indigenous disease-transmitting insects and pathogens have been added to those introduced from elsewhere – such as foot-and-mouth disease of cattle. Many tropical cattle diseases are transmitted by ticks with two or more hosts, making control and elimination particularly difficult. Among the most contagious and serious are rinderpest, a viral disease which has spread from water buffalo to domestic cattle, and trypanosomiasis (or nagana), one of a wide range of diseases of both humans and domestic livestock transmitted by various species of the tsetse fly (*Glossina* spp.). Three groups of flies associated with different habitats are recognized: *G. fusca* or forest flies; *G. morsitans* of the savanna; and *G. palepals*, which unlike the other groups can live in close association with man, particularly where shifting agriculture causes renewal of the thicket type of vegetation which is the essential breeding habitat of the tsetse. Elsewhere dense agricultural populations tend to clear and eliminate woodland completely. While there is little doubt that the tsetse is in retreat as forest is cleared for agriculture, there are still very extensive areas, particularly in Africa and South America, where its presence precludes all but the most disease-resistant cattle such as the native N'dama and Mutura breeds of West and Central Africa respectively. The control and elimination of disease in livestock is made particularly difficult by the presence in the Tropics – particularly in Africa – of remaining large populations of wildlife with habitats and niches similar to those used by domestic livestock and which provide natural breeding reservoirs for pests and pathogens.

SOIL POTENTIAL

It has become increasingly evident that the soils of the humid Tropics are neither as agriculturally fertile as the biomass and productivity of the natural tropical vegetation might suggest nor as uniformly infertile as their yields and agricultural condition has led one to assume. The nutrient status and agricultural potential of soils associated with a humid tropical climate are, in fact, very varied depending on the parent material from which they have been derived; the degree of weathering and leaching to which they have been subjected; the prevailing climatic conditions; and, not least, the way in which they have been managed. They are, however, difficult soils to cultivate and pose problems different in either kind and/or intensity from those in temperate areas. This is partly a result of their inherent characteristics and partly because of the climatic conditions under which they occur (see Tables 10.3 and 10.4).

The most extensive and characteristic of the tropical soils are the latosols – a generic term for a wide variety of red and yellow freely drained soils (Young, 1974b). Those developed on acid parent materials (acid igneous rocks, sandstones etc.) have, to a greater or lesser degree, been highly weathered and intensively leached, and consequently lack a reserve of weatherable minerals in the soil profile. The clay component is usually dominated by kaolinite and amorphous iron and aluminium oxides; and the cation exchange capacity and water-holding properties of the soil are low. Indeed, despite the high clay content of many tropical soils, *micro-aggregation*, as a result of cementation by iron oxides, results in a fine 'sand-like' granular structure which facilitates free drainage and ease of cultivation. Generally the latosols are moderately to highly acid in reaction, strongly buffered and nutrient-deficient, particularly in calcium and phosphorus. The latter is, in the acid soil environment, very susceptible to fixation in an unavailable mineral form. In addition, aluminium becomes mobile and attains levels of toxicity which inhibit or prevent the growth of all but the most aluminium-tolerant crops.

Table 10.3 Relations of latosol types to environmental factors

	Parent material		
Climate	*Acidic (siliceous or intermediate)*	*Acidic (highly weathered)*	*Basic (rich in ferromagnesian minerals)*
Rain forest (mean annual rainfall >1500 mm)	leached ferrallitic soils		basisols: strongly leached ferrisols* (c. 1800 mm)
Savanna (mean annual rainfall 600–1500 mm)	ferruginous soils	weathered ferrallitic soils	ferrisols (*c.* 900 mm) eutrophic brown soils
Tropical high-altitude (mean annual rainfall >1500 mm)	humic ferrallitic soils humic latosols		humic ferrisols

* Land-form factor important: gently undulating erosion surfaces.

(adapted from Young, 1974a)

The poorest of the latosols are the *weathered ferralites* or *plateau latosols* characteristic of the old erosion surfaces of the high plateaus of the savannas in Africa and South America. They are weakly structured or massive, often with a 'sandy texture'. Once exposed by cultivation, they become hard and cement-like at the end of the dry season. The best agriculturally endowed latosols are the basisols – those derived from base-rich parent material – particularly basic and ultra-basic igneous rocks rich in ferromagnesian minerals. In this case the mineral soil may contain a reasonable amount of weatherable minerals which provide a nutrient reserve and the clay component has a high proportion of montmorillonite. As a result cation exchange capacity and water retention is higher than in the poorer latosols and leaching is not so intense. While the plateau ferralites are frequently left uncultivated and remain under a poor grassy savanna-type vegetation, the basisols are the most productive and are often used for the more nutrient-demanding commercial monocultures of tea, coffee and cotton for example. In all the latosols, level of organic matter, and consequently nitrogen content, is low. The amount in uncultivated soils is a function of the type of vegetation and annual volume of litter production, i.e. 3–5 per cent in lowland humid rainforest; 2–3 per cent in moist savanna; 1–2 per cent in dry savanna.

Lateritic horizons (sometimes referred to as ironstone, ferricrete, murrain) are of frequent occurrence in most types of latosols – in *c.* 5–10 per cent of the soils of the

Table 10.4 Chemical and physical characteristics of the main types of latosols

	Ferrasols			Ferrisols
	Ferruginous	Ferrallitic	Weathered ferrallitic	Basisols (most common ferrisols)
Depth to weathered rock	2–5 m	>10 m	very deep (lateritic horizon common)	deep
Percentage organic matter	1–2%	2–5%	low	1–2%
pH	5.5–6.5	<5.5	–	7.5–8.5%
Base saturation	40–90%	<20.0%	–	–
Cation exchange capacity (me = milli-equivalents)	25 me 100 g^{-1}	<20 me 100 g^{-1}	–	–
Clay mineral	moderately strongly leached kaolinite with some illite and montmorillonite	dominated by kaolinite with free Al, Fe	kaolinite	rich in ferromagnesium

(collated from data from in Young, 1976)

humid Tropics according to Young (1974). They can vary in thickness from 3 m to, exceptionally, 30 m. Composed mainly of cemented iron and aluminium oxides varying from 80–90 per cent Fe_2O_3 to over 60 per cent Al_2O_3 (bauxite), laterite is normally very hard – indeed more like rock than soil. It can be massive with a cellular or vesicular structure; nodular with varying degrees of cementation; or be ferruginized rock. Occasionally, limited deposits of soft laterite occur which harden on exposure to air, but this form, contrary to former belief, is the exception rather than the rule. Lateritic horizons can seriously limit soil depth, and restrict root penetration; soils above them rapidly become water-deficient in the dry and saturated in the wet season. Where they lie at shallow depths under relatively level land with poor surface drainage, they can cause extensive flooding.

Soils other than latosols and of higher agricultural value are, in comparison, relatively limited in extent in the humid Tropics. Two of the most agronomically significant are the *vertisols* and the *alluvial* soils. The *vertisols* (originally named *grumusols*) are dark-coloured 'cracking clays' variously referred to as black cotton, or black earths, of the regur soils on the Deccan of India, and as gilgai clays in Australia. In the Tropics they are mainly associated with wet and dry climates and natural vegetation varying from semi-deciduous forest to savanna grassland. A high clay content gives them a relatively high water-holding capacity but permeability and water conductivity are very low indeed. Hence leaching is restricted and calcium and magnesium are the dominant cations with sodium often abundant. Concretions of $CaCO_3$ are present in greater or lesser amounts throughout the profile. Soil reaction tends to be slightly alkaline, with pH 7.5–8.5. The range of soil consistency is narrow: very hard when dry, passing rapidly to a plastic state when wet and really only becoming friable when a surface mulch develops.

The diagnostic characteristic of the vertisols is a texture dominated by montmorillonitic clays (40–90 per cent) whose extreme swelling and shrinkage properties can cause self-mulching of the soil surface. As they dry out, these soils become hard and cohesive, and shrinkage results in the formation of deep vertical cracks and fissures. Once initiated, these facilitate further drying with a concomitant widening and deepening of the fissures. At the same time the surface soil disintegrates into a granular structure or 'mulch', which may attain a thickness of 2 cm and seal the soil surface and fill in the cracks. During the rainy season the uneven absorption of water by the mulched material, particularly in deep cracks and fissures, causes differential swelling and associated vertical and horizontal soil movement. This results in the formation of an irregular hummocky surface and slickenside fractures within the soil, which impede rather than facilitate water movement.

The most productive soils in the Tropics (capable of supporting the highest densities of agricultural populations in the world) are undoubtedly the *alluvial* soils of the great riverine lowlands of Monsoon Asia. Their agricultural value and capability are closely related, on the one hand, to the physical and chemical composition of the material of which they are formed and, on the other, to the annual deposition of a fresh load of alluvium during the height of the wet season when extensive inundation of flood plains takes place. Their use and yield potential are closely related to the length of time they are inundated and the depth to which they are saturated. They are the basis of paddy rice cultivation in Asia and will be discussed in greater detail in the next chapter.

As important as the inherently low fertility of many tropical soils is the difficulty of either maintaining or increasing fertility. Under natural conditions most of the nutrients in the ecosystem tend to be concentrated in the above-ground plant biomass; and rapid decomposition under humid climatic conditions results in a smaller amount of organic matter in the soil than in humid temperate regions. Also, because of the low cation exchange capacity of the clay minerals of many soils, the available nutrient reservoir in tropical soils is highly concentrated in the small amount of organic matter contained largely in the top 20 cm of the profile. Cultivation opens up and lays bare the soil to the full impact of atmospheric conditions. Surface and subsurface temperatures increase and rates of organic decomposition are speeded up. Indeed soil temperatures of 40 °C are common and can occasionally reach 50 °C. Over 35 °C, however, the temperature becomes supra-optimal for crop growth and yields are consequently depressed.

Removal of the former vegetation cover reduces interception losses, thereby increasing the already high intensity of leaching or conversely the amount of surface run-off. The impact of high-intensity rainfall on bare soil causes the breakdown of surface structures, the sealing of micropores and the formation of surface crusts by the fine mineral particles so produced. Many of the poorer latosols and the vertisols

Table 10.5a   Annual run-off as a percentage of rainfall at Kwadaso and Ejura

	Run-off (%)		
	1974	*1975*	*1976*
3.0% slope			
Bare fallow	35.8	38.0	36.4
Mixed cropping	4.6	3.7	5.1
7.5% slope			
Bare fallow	48.0	34.0	49.8
Mixed cropping	12.4	14.9	13.2

(from Aina *et al*, 1979)

Table 10.5b   Soil loss at Kwadaso and Ejura

	Soil loss (tonne ha^{-1} yr^{-1})		
3.0% slope			
Bare fallow	18.3	19.7	18.3
Mixed cropping	2.5	1.3	2.5
7.5% slope			
Bare fallow	100.1	187.1	313.0
Mixed cropping	2.3	20.3	33.5

(from Aina *et al*, 1979)

are liable to become very hard and impermeable as a result of either 'sun-baking' and/or compaction. Crops can be lost if this happens before harvest. Soil degradation as a result of depleted organic matter and loss of surface structure gives rise to soil erosion hazards much greater in the humid Tropics than in temperate regions because of the much higher erosivity of tropical rainfall (Table 10.5a, b). Further, the resulting loss of the same amount of top-soil will cause a proportionately greater depletion of soil nutrients in the Tropics and is particularly serious for shallow 'gravelly' soils. Lal (1984) notes that a cumulative soil loss of 50 tonne ha^{-1} results in yield reductions in maize of 56 per cent on slopes of 15 per cent; 14 per cent on slopes of 10 per cent; and 8 per cent on slopes of 1–5 per cent. Lal and Greenland (1979) suggest that, in the majority of latosols in the humid Tropics, cultivation is in fact deleterious for crop production and that no-tillage systems combined with mulching are methods better suited to soil conservation in these areas.

NUTRIENT CYCLES AND MAINTENANCE OF SOIL FERTILITY

The production of annual food crops in the humid Tropics is achieved largely without the input of either organic or inorganic fertilizers. The production of fodder crops and the use of animal manure is exceptional and there is little real integration of livestock keeping and arable cultivation. Where animals (cattle, goats, sheep) are kraaled regularly the dung is usually reserved for the vegetable garden or more often is used for domestic fuel, particularly where there is a paucity of fuelwood. In the still forested and wooded areas of the very humid Tropics the tsetse fly precludes the keeping of cattle. In the semi-humid savanna areas livestock range widely over uncultivated areas and their dung is hence dispersed and its impact on soil fertility considerably reduced. In addition, the cost of mineral fertilizers and other inputs is prohibitive for the majority of subsistence farmers. The effectiveness of mineral fertilizers alone can be marginal because of high weed competition and intensity of leaching. Maintenance of soil fertility is therefore dependent on a management system which alternates a period of soil cropping and nutrient depletion with one of soil resting and natural nutrient build-up. This is the basis of the systems of *fallow cultivation* widely practised throughout the humid and dry Tropics (Ruthenberg, 1980).

A distinction is usually made between extensive and intensive fallow systems, the former represented by *shifting cultivation*, the latter by *bush* and *savanna* (or *grass*) *fallows*.

SHIFTING CULTIVATION

Shifting cultivation is still the most extensively practised system of farming in the humid rainforest environments of South America, Africa and South-East Asia. Wrigley (1981) gives a figure of 30 per cent of the cultivable soils in the Tropics. A clearing is cut in the forest; the felled trees are burned in the dry season. The ashes provide a store of readily available nutrients, and crops are planted with the aid of a digging stick and minimum soil disturbance at the beginning of the rainy season. A mixture of food crops with varying growth periods, nutrient requirements and food value are planted – including maize or rice, squash, cassava, bananas, beans and a wide variety of vegetables. This ensures a continuous supply of food throughout the year-long growing season as well as a complete and protective ground-cover. After about 3 years cropping, yields begin to decline and tree growth starts to encroach. The plot is abandoned and the whole farming community or tribe moves to clear and cultivate another part of the forest. Reforestation of the abandoned land is rapid, with trees of up to 15 m high within 10 years, the length of time it is thought necessary to restore the nutrient status depleted by 3 years cropping, though the length of time between successive cultivations of the same land area varies with population density.

FALLOW CULTIVATION

Fallow cultivation *per se* differs from shifting cultivation in that tillage is by hoe and the areas cultivated are usually larger. The length of time land is left fallow is shorter and the farming community is sedentary. Fallows of varying length and vegetation occur throughout the humid and subhumid Tropics. A broad distinction is usually made between *bush fallow*, which is more characteristic of humid to semi-humid areas and which, particularly in West Africa and South-East Asia, is often associated with a tree crop; and *grass fallow*, which is predominant particularly in the African savanna environment. In contrast to shifting agriculture, the fallow vegetation is burnt at the end of the dry season and planting starts after the rains commence when the soil is sufficiently moist. Under a drier and more seasonal climatic regime, crop combinations shift to sorghum and/or millet, yams, and crop legumes (groundnuts, soya beans or chick peas). To give sufficient depth for tuber development and to avoid competition from the persistent roots of perennial savanna grasses, crops are planted on mounds or ridges. In the drier grassy tsetse-free savanna areas crop stubble and fallow land is normally grazed either by community livestock or more commonly by nomadic or semi-nomadic pastoralists. The length of either bush or grass fallow varies within tribal land areas according to soil potential and/or accessibility from the village. In the former instance the better soils are used more intensively on a short-fallow rotation, the poorer on a longer extensive fallow system. In the latter the land nearer to the settlement will be more intensively managed than that at some distance from it.

The length of fallow is also determined, to an increasing extent, by population density: throughout the Tropics, particularly in Africa and in South-East Asia, population increase has been accompanied by a reduction in the length of fallows. In Gambia the long bush fallow which formerly was of 30 years duration has contracted in three decades to 3 years. As the fallow shortens, so re-establishment of park savanna is inhibited and grass fallow inevitably replaces a former bush fallow. In the drier grassland areas the fallow rest is insufficient to make good the combined losses from the system by cropping and grazing. At best the soil can be maintained at a very low level of productivity; more commonly it is subject to soil degradation and rapidly becomes susceptible to accelerated soil erosion.

Weed infestation and competition is particularly characteristic of all fallow systems, where it constitutes the most pervasive yield depressant. Weeds which get a foothold in cultivated land are maintained and proliferate in fallow vegetation as fertility declines. Several authors, however, suggest that weeds may perform a positive as well as negative role in maintaining fertility and reducing erosion: their shade and litter may help to protect the soil from direct insolation. Many weeds are deep-rooted and can recycle nutrients from depth; others are valuable nitrogen-fixers. Further, the diversity of plant species, to which weeds contribute, helps to maintain a balance between insect pests and their predators better than in weed-free monocultures.

CONTINUOUS CULTIVATION

Continuous permanent upland (i.e. without irrigation) cultivation, with or without a short between-harvest fallow, is more common in areas of high population and/or of high rainfall, an extended rainy season and soils with good water-storage properties. It has long been characteristic of intensive cultivation of tropical vegetable gardens and Chinese rice cultivation, where there is a high input of organic wastes including sewage (or 'night soil'), or where, as in parts of South-East Asia, soils derived from nutrient-rich volcanic soils can produce two or more crops per year. Permanent agriculture is also characteristic of areas of dense population, particularly in India, and is increasing as pressure on limited land resources grows. Maintenance of agricultural production, without input of nutrients and in the absence of a fertility-restoring fallow period, is difficult. It has, inevitably, to be more dependent on carefully balanced crop rotation systems.

CROP COMBINATIONS

The diversity of crops produced from a given land area during the growing season is a characteristic feature of much tropical agriculture. One tribe in Zaire was reported to cultivate 80 varieties of 30 species of food crop (Okigbo and Greenland, 1976). The aim of the majority of farms in the humid Tropics is to produce as large and as reliable a source of food as possible. To this end a mixture of food crops which ensure the highest return possible from a given amount of land during the growing season is cultivated. Another objective is to reduce the risks of crop failure associated with climatic variability and the proliferation of crop pests and pathogens by the use of drought- and/or disease-resistant varieties, together with methods of delayed and/or phased planting. These aims will be more successfully achieved if cropping methods help to minimize the high risk of soil erosion by providing continuous protective cover, to maximize the use of soil nutrients and water and to conserve soil fertility by crop rotation.

Three biophysical factors combine to influence the type of farming: regional climate, soil conditions and type of crop. The dominant combinations which characterize the more humid Tropics are: maize and/or rice; sweet potatoes, bananas, squash, cassava and yams; and tree crops (oil palm, cocoa, rubber). With increasing length of dry season the scenario shifts to sorghum and millet; yams and/or cassava; and crop legumes such as groundnuts, chick peas, soya beans, field beans etc. Within the broad climatic framework the particular crop system may vary with the physical or chemical characteristics of the soil. Heavier more water-retentive soils determine the balance between yams and the more drought-resistant cassava, while more demanding crops like maize and sorghum are grown on soils of higher nutrient status than is millet. In addition, the phenological characteristics of the growth form and environmental requirements of the particular crop or crop variety all determine the compatibility of cultivars and their ability to form successful mixtures or crop associations. Further, crop combinations may be a function of the stage in the development of the cropping system, or of the relative importance of cash and food crops of tree and field crops respectively. Successful crop combinations are those in which crops make varying but complementary demands on the soil and perform varying but complementary roles, or occupy varying habitats and/or niches within the particular agro-ecosystem.

The most important staples are the *cereals*, sorghum and millet, whose temperature requirements can be satisfied only within the Tropics, and maize and rice, adapted to a wider range of temperatures (see Table 10.6). Maize and rice have high water needs during their growing periods. Both are intolerant of soil water deficits,

Table 10.6   Principal cereal crops in four rainfall zones of West Africa

Region	Mean annual rainfall (mm)	Length growing season (days)	Main crops	Secondary crops
A	350	60	SS millet	–
B	550	100	SS, MS sorghum MS millet	SS sorghum SS millet
C	800	120	MS, LS sorghum MS, LS millet	SS maize SS sorghum SS millet
D	1000	145	LS maize LS sorghum LS millet	SS maize SS sorghum SS millet

SS, MS, LS = short-, mid- and long-season cultivars.

(from Norman *et al*, 1984)

while maize performs best where high temperatures are associated with high atmospheric humidity. Sorghum and millet, with shorter growing periods, need less water and can tolerate high temperatures, while sorghum is better adapted to drought than millet. The latter is, however, better able to produce a harvest with much less precipitation and on much poorer soils than sorghum.

The cereals have relatively short determinate growth cycles. They are often the first crop to be planted because they mature early, can take advantage of an initial but sometimes short-lived rainy spell and are more demanding of nutrients than many other crops; or they can be grown as a catch crop towards the end of the rainy season. The tall erect habit of cereal plants provides, as in the case of sorghum–yam mixtures, a useful support for the climbing crop plant. In addition, sufficient soil moisture allows continuous tillering of sorghum, after the grain harvest; this produces a *ratoon* crop of straw which can be cut for a variety of uses including animal feed or which can be grazed.

In parts of the very humid Tropics, where a marked dry season is absent, cereals are displaced as the most important staples by *bananas* or *tuberous root crops* such as yams and cassava which have a year-long growing season (see Fig. 10.7). The banana is one of the most important non-cereal food and export crops of the humid Tropics. It is easy to establish; it yields fruit 12 months after planting and continues to produce all year round. With a great number of varieties, it is adapted to a wide range of soil conditions, provided they are well drained. Climatic requirements include mean monthly temperatures of not less than 15 °C combined with high precipitation and high atmospheric humidity. Yams (of which there are twelve species of high food value) and cassava need high rainfall and a long growing season. The former take between 6 and 12 months to mature, and can either be harvested at the end of the wet season or be left in the ground for up to 120 days without serious

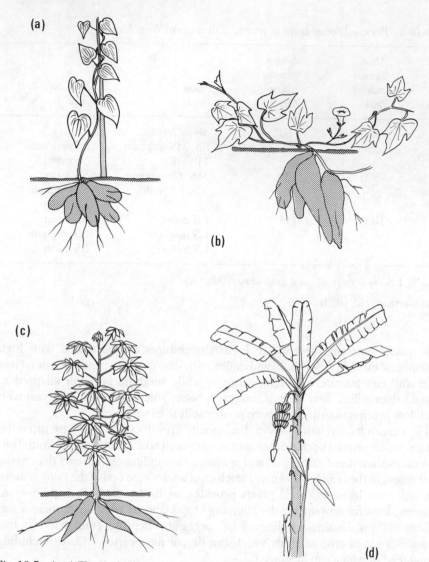

Fig. 10.7 (a–c) Tropical tuberous crops: (a) yam, (b) sweet potato, (c) cassava; (d) banana (adapted from Norman *et al*, 1984)

weight loss or deterioration. Cassava, whose tubers mature in 6–12 months or more, depending on variety, has a wider climatic range than the yam. It can grow from sea level to 1800 m in the Tropics where mean annual rainfall is over 5000 mm. In addition, a long and well-developed root system makes it capable of tolerating or evading drought and producing a crop on soils of very low nutrient status.

The third important group of tropical food crops which provide a complementary source of energy and, particularly, of protein and fats are the *crop legumes* (see Table 10.7). These comprise a very large number of species and varieties of which the most important are the common bean, the groundnut (or peanut) and the soya bean.

Table 10.7   Comparative protein yield of cereals and legume crops

Crop	Average yield (tonne ha⁻¹)	Protein content (%)	Average protein yield (kg ha⁻¹)
Cereals			
Rice (paddy)	2.01	7.5	153
Maize	2.24	9.5	118
Sorghum	0.83	10.5	87
Millet	0.55	10.5	58
Legumes			
Soya bean	1.33	38.0	505
Groundnut	0.85	25.5	217
Beans	0.60	22.0	132
Chick pea	0.66	20.0	132

(from Norman *et al*, 1984)

They all have varieties whose pod-maturing period varies from 2 to as much as 10 months; all require at least 500–600 mm of precipitation during the growing season; all can tolerate a wide temperature range and an extratropical optimum. Most, however, are intolerant of cold and cannot survive frost, although several can tolerate high temperatures and drought. The crop legumes perform a diversity of roles within tropical agro-ecosystems, not least that of soil conservation. Many varieties are either low-growing or climbing and can be integrated with the structure of a cereal crop and can utilize soil water after cereals are harvested. Their prolific vegetation can effectively suppress weeds and protect the soil surface from erosion. Finally, all legumes fix nitrogen (see Table 10.8). However, the amount derived

Table 10.8   Nitrogen fixed by tropical annual legumes under field conditions

	Nitrogen fixed (kg ha⁻¹)
Cajanus cajan (pigeon pea)	224
Vigna ungnicalata (cow pea)	198
Cyamopsis tetragonoloba (cluster bean)	130
Arachis hypogaea (groundnut)	124
Glycine max (soya bean)	103
Cicer arietinum (chick pea)	103
Vigna radiata (green gram)	61

(from Norman *et al*, 1984)

from this source by other plants is not known exactly. It has been suggested that competition for nitrogen is less when a legume and non-legume crop are grown together. Also, the residual effect of legumes on subsequent cereal crops has been demonstrated, uptake being twice as great as when no preceding legume has been cultivated.

Table 10.9   Growth requirements and duration of principal perennial crops of the humid Tropics

Perennial field crops	Climatic requirements	Years to first crop	Years of production
Sugar cane	m.m.t. 30–32 °C ppt. 900 mm + high sunshine	1.0–1.5	4–6
Banana (for export)		1.0–2.0	5–5.5
Pineapple		1.5	3.5
Sisal		3	8
Shrub crops			
Coffee	m.m.t. coolest 11 °C max. 30 °C ppt. 900 mm–200 mm frost-intolerant	3	12–50
Tea		3	50
Tree crops			
Oil palm		3–4	35
Rubber		2–4	35
Cocoa	m.m.t. not less 10 °C ppt. 900 mm	8–11	8–100
Coconut		4–6	100
Cloves		8–9	100

m.m.t. = mean monthly temperature; ppt. = precipitation.

(from Ruthenberg, 1976)

Most of the tropical food crops, with the exception of the banana and, sometimes, cassava, are either true annuals or are cultivated as annuals. The long-run perennials are largely non-food crops, and frequently important commercial crops are *trees* and *shrubs* (see Table 10.9). The most commonly cultivated tree crops of the Tropics require constantly high temperatures and abundant moisture during a relatively long growing season. Their potential range lies within the more humid Tropics (in the rainforest zone of which they are natural components) and include rubber, oil palm, cocoa and coconut as well as the evergreen shrubs, tea and coffee. The trees form important multiple-use components in crop mixtures, particularly in West Africa. A large evergreen LAI ensures high primary productivity; and harvestable yields are more reliable and consistent than from the annual food crops. Once established, tree crops require minimum soil cultivation and weeding. Also, they effectively protect the soil surface, helping to reduce leaching and erosion, while at the same time recycling a high proportion of the nutrients extracted from the soil via their leaf fall. Finally, they lend themselves to the establishment of a stratified understorey cultivation of crops of varying growth height which replicates that of the natural forest. Thus the tree crop is widely regarded as the type of crop best adapted to produce and sustain high yields in the humid Tropics.

Tea and coffee are trees managed as shrubs. Both have high water needs but can tolerate a wider range of temperature from tropical to subtropical provided there is no risk of frost. Both are cultivated at higher and cooler altitudes than the other tropical crops: tea at 1500 m and over; coffee at up to 700 m. However, in comparison to other tree crops, labour demands for weeding, pruning and harvesting are very high.

Monocultures of field crops, or of plantations of shrub and tree crops in the humid Tropics are mainly associated with tea and coffee; sugar; cotton and sisal; bananas; and the commercial production of vegetable oil, cocoa and pineapple. While the percentage of the total cultivated acreage under these cash crops in the Tropics is small, yields are high. Management, however, is intensive with high inputs of inorganic fertilizers; of water for irrigation, particularly for cotton and sugar; and of herbicides and pesticides. The monoculture is, however, particularly susceptible to pest infestation in the Tropics. Ruthenberg (1980) notes that the commercial field crops were the first of the tropical products to benefit from technical innovations in methods of cultivation and from plant breeding for high-yielding disease-resistant varieties, the impact of which has been most marked in the case of sugar, pineapple and cotton. Yields in these crops have doubled within the last 30 years. While commercial tree crops provide a continuous or near-continuous plant cover and a high organic matter content in the soil, maintenance of high yield is at the cost of comparatively higher input of nutrients than in temperate intensive agriculture and an even greater risk of yield reduction or crop destruction by pests and disease.

Chapter 11

Rice

Rice is the major food crop of the Tropics, particularly of South and South-East Asia, which accounts for c. 50 per cent of the total world acreage and production. Although most of the world's rice is produced within the Tropics, the actual and indeed potential range of the crop extends both north (45°N) and south (40°S) into warm- and cool-temperate climatic regimes. Rice is adapted to a wide range of environmental conditions: to average air temperatures of 20–38 °C and optimum soil temperatures of 25–35 °C during growth (Table 11.1); to varying intensities and duration of light; and to moderately acid to moderately alkaline soils, with low to high nutrient status and of varying textural characteristics. It is, however, the least drought-resistant of all the cereal crops and soil water availability is the main environmental factor limiting both distribution and yield of rice.

Table 11.1 Effect of temperature on various processes of rice growth

	Critical temperature (°C)		
Growth stage	Low	High	Medium
Germination	16–19	45	18–40
Seeding emergence and establishment	12–13	35	25–30
Booting	16	35	25–28
Leaf elongation	7–12	45	31
Tillering	9–16	33	25–31
Initiation of panicle primordia	15	–	–
Panicle differentiation	15–20	30	–
Anthesis	22	35–36	30–33
Ripening	12–18	>30	20–29

(from de Datta, 1981)

The extensive geographical range of rice is a function of its tremendous genetic potential. Early domesticated and long cultivated in South-East Asia, the rice cultigen comprises a vast array of cultivars adapted to a diversity of physical habitats and methods of cultivation (de Datta, 1981). The IRRI has recorded nearly 1500 varieties, the perpetuation of which may be a result of traditional methods of harvesting involving ear-by-ear collection with implements which cut the rice plant singly. Varieties differ in form, habit and physical behaviour. Some are tall, others short; some tiller more than others; some are more susceptible to lodging or to ear-shattering than others. Varieties also differ in the length of their maturation period, from a short (90–100 days) to long (>200 days) growing period (Fig. 11.1); in their soil water requirements; and in photoperiodic sensitivity from short-day to long-day. Finally, rice cultivars differ in their tolerance of, or resistance to, adverse environmental conditions such as drought, flood, high wind force, soil salinity, and other soil toxicants (iron, aluminium) or deficiencies (zinc).

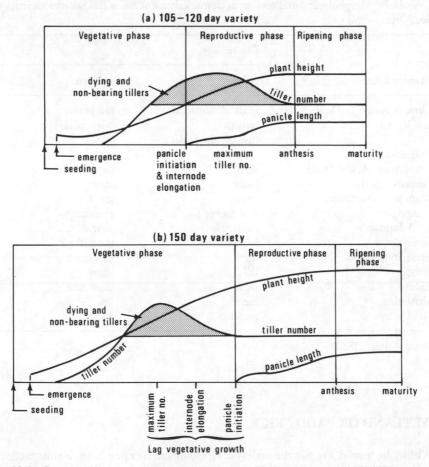

Fig. 11.1 Rice: development of cultivars sensitive to (a) short duration (105–120 days); (b) long duration (150 days) (from Yoshida, 1977)

185

There are only two species of cultivated rice: *Oryza sativa*, which is predominant, and *O. glaberrima*, which is only grown in very limited areas of West Africa. Within *O. sativa* three major groups of traditional varieties have long been distinguished on the basis of their geographical distribution:

1. *Indica*.
2. *Japonica*.
3. *Javannica*.

The contrasting characteristics of the *indica* and *japonica* varieties are summarized in Table 11.2. Cutting across these traditional groups are the more recently developed HYVs, the products of inter- or intragroup selection and breeding. These have replaced the older varieties in certain parts of the Asian heartland and dominate the more recently established rice crops that are grown in the extratropical periphery.

Table 11.2 Morphological and ecological characteristics of *indica* and *japonica* varieties of rice (*Oryza sativa*)

	Rice variety	
Characteristics	*indica*	*japonica*
Climatic zone	tropical monsoon	temperate
Day length	sensitive (usually short-day)	not sensitive
Tolerance to unfavourable conditions	high	moderate
Vegetative period	long	short
Response to fertilizers	poor	good
Lodging	susceptible to	resistant to
Seed dormancy	present	absent
Grain shape	long/narrow	short/thick
Diversity of forms	high	moderate
Height	tall	short
Tillers	many	moderate
Shattering	high	low
Yield potential	medium	high

(from Wrigley, 1981)

WETLAND OR PADDY RICE

Within the humid Tropics the variety of rice cultivated depends on the interaction of three major variables: environmental conditions, cultural traditions and whether production is for subsistence food or for sale. Within the humid Tropics of South and South-East Asia *c.* 90 per cent of the rice is cultivated under wetland conditions where the soil within the root zone must remain saturated for most of the growing period. Rice is not a true (i.e. *obligate*) aquatic species. It can, however, grow well in saturated soil because of the morphological adaptations of its shoot and root system. As well as being shallow and laterally spreading, the roots have pore spaces that can hold and retain air. In addition, the plant's respiratory system is adapted to a lower oxygen availability, while the roots can effect oxidation in the rhizosphere, both of which processes help to compensate for the anaerobic conditions associated with soil saturation.

The dominance and concentration of *wetland* (or *paddy*) *rice* in Asia is thought to be a result of a fortuitous combination of a tropical monsoon climatic regime, characterized by very high precipitation with a marked seasonal distribution and a well-defined dry season, with the very considerable extent of low-lying level flood-plains and deltas, which lend themselves to labour-intensive wetland agriculture. Moormann and van Breeman (1978) have categorized wetland cultivated for rice according to the main source of the soil water supply as: *phreatic* (dependent on ground water) and *fluxial* (dependent on surface run-off) (see Table 11.3).

While in many instances it is not easy to assess precisely the relative contribution from each of these sources, in others one or other becomes dominant. Most of the Asian paddy land is dependent primarily on direct precipitation (80 per cent according to Grist, 1976) and surface run-off from upper to lower slopes. Cultivation of paddy rice is dependent on a complex, indeed highly sophisticated, system of water control by the levelling and bunding of small land areas in order to facilitate the collection and retention (i.e. ponding) of surface water at the optimum depth for successful rice cultivation. The latter is difficult to define and averages of up to 20 cm have been quoted. Research by the IRRI indicates that high efficiency of water use and good yields are achieved with a continuous flow of shallow water (i.e. 2.5 cm) over a typical clay soil. Under these conditions soil and water temperatures are higher during the day and lower at night than in deeper water; and decomposition of organic matter is rapid, resulting in vigorous root development. However, there is no standard practice since local sites and crop types vary so greatly, while mismanagement and/or fear of drought during the growing season may result in a more extravagant use of water than is necessary. Ponding also ensures maximum infiltration and percolation and the creation of a subsurface soil layer 10–30 cm deep, which is frequently perched some metres above the regional water-table. The development and depth of the saturated zone depends on the rate of infiltration/percolation plus the water-holding capacity of the soil. Soil with a percolation rate of $1 \, \text{mm day}^{-1}$ is ideal. In many paddy fields it is $5–10 \, \text{mm day}^{-1}$, at which rate insufficient water is held to ensure a second crop. Good paddy requires heavy soils, i.e. preferably those with over 70 per cent fines (silt plus clay) which are characterized by slow hydraulic conductivity, low permeability and an impervious layer at shallow depth and/or cultivation techniques which will help to produce and maintain these physical conditions in the soil.

Paddy cultivation is unique in that the main method of soil tillage is by *puddling* or tillage of saturated soil with standing water. This brings about the partial or complete destruction of soil aggregates and a consequent reduction in the infilt-

Table 11.3 Terminology of rice lands as a function of physiography and hydrology

Intensity of irrigation	Physiographic hydrologic category	Bunding and levelling	Flooding regimes*	Terminology proposed	
Zero or low: availability of water dependent on supply	pluvial	without	no flooding	pluvial rice land	dry-land
		with	1,2, (3)	pluvial-anthraquic rice land	wet-land
	phreatic	without	no flooding	phreatic rice land	wet-land
		with	1,2,3	phreatic-anthraquic rice land	wet-land
	fluxial	with or without	2,3,5,6, 7,8	fluxial rice land	swamp-land
High: available independent of natural supply in 4 out of 5 years	irrigated	rarely without	4	irrigated rice land	irrigated

* Regime 1: shallow, irregular, brief flooding; regime 2: shallow, irregular, prolonged flooding; regime 3: shallow, continuous, uncontrolled flooding; regime 4: shallow, continuous flooding controlled by irrigation; regime 5: shallow to moderately deep seasonal flooding; regime 6: deep seasonal flooding; regime 7: moderately deep to shallow flooding after recession of deep floods; regime 8: tidal flooding.

(from Moormann and van Breeman, 1978)

ration/percolation rate, an increase in the water- and nutrient-holding capacity of the soil and suppression of weeds. Puddling is effected either by hand-tools or by ox-drawn ploughs. Over a long period of cultivation it is usually accompanied by the formation of a temporary (seasonal) or permanent compacted *traffic pan* some 5–10 cm thick at 10–40 cm below the surface. In Japan it has been estimated that incipient pans appear after 50 years cultivation; fully developed permanent pans after 200 years. The pan further inhibits percolation and can facilitate surface ponding even on initially well-drained soils. It also allows permeable soils which lack a high water-table to retain water and serves to make wetland more accessible to the farmer and his draught animals. If the pan is disrupted, soils can become muddy and non-resistant to pressure to a considerable depth, a hazard which limits the use of heavy vehicles such as tractors.

Flooding and soil saturation help to combat the proliferation of weeds, to which rice cultivation is peculiarly subject. Of the ten weeds which cause 90 per cent of the world's food-crop losses by competition, three are common on paddy land: *Cyperus rotundus, Echinochloa crus-galli* and *E. colonum*; and the water hyacinth (*Eichhorria crassipes*) can block irrigation ditches and canals. More importantly, however, saturation initiates chemical changes in the soil on which the successful nutrition of the rice crop depends (see Fig. 11.2). After flooding, the oxygen content of the soil is rapidly depleted and becomes restricted to a shallow (*c.* 0.5–10 cm) surface layer (the oxidized zone) into which atmospheric oxygen can diffuse easily. Below this zone reduced anaerobic conditions prevail. Rice roots, however, can continue to respire aerobically because of oxygen penetration via the air spaces and channels in the shoots and roots. Lateral growth results in a higher concentration of roots at shallow depth, hence nutrients need to be available within 20 cm of the surface for maximum growth rates.

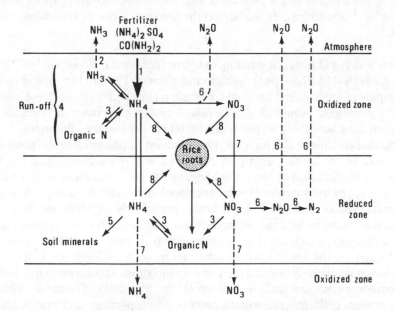

Fig. 11.2 Fate of nitrogen fertilizers in wet-rice soil: 1, urea hydrolysis; 2, ammonia volatilization; 3, nitrogen immobilization; 4, run-off; 5, ammonium fixation; 6, denitrification; 7, leaching; 8, plant uptake (from Craswell and Vlek, 1979)

Reducing conditions dominate the saturated zone. Ferric iron and manganese oxides become ferrous and mobile; sulphates and nitrates are also reduced, while incomplete decomposition (fermentation) produces CO_2, CH_4, H_2S and organic acids. More significantly, however, the availability of phosphorus, silica and molybdenum and the supply and availability of nitrogen are increased. Nitrogen fixation more than compensates for the losses resulting from denitrification from the reduced soil zone. Nitrogen availability to wetland rice is greater than to rice grown under non-saturated conditions on the same type of soil (Russell, 1973). In the former conditions nitrogen fixation is carried out by four main groups of organisms for which paddy land, with warm clear relatively still water and high amounts of sunshine, provides an ideal habitat. They are:

1. *Blue-green algae* (*Tolypothrix, Nostoc, Schizothrie* and *Calothrix* spp.), which inhabit the water above the soil surface and can fix over $70 \, kg \, ha^{-1}$.
2. *Azolla pinnata* (the floating fern), which contains a blue-green algae symbiont, *anaebena azolla*, and fixes $c. \, 70–110 \, kg \, ha^{-1}$.

   Phosphorus fertilizer can stimulate the growth of these plants and increase the nitrogen yield in both groups. They develop either on the wet soil surface or in the soil and rise to the water surface where they continue to grow. Pulled down as the rice plants grow or when surface water is depleted, they decay on the soil finally to be incorporated within it by puddling.
3. *Heterotrophic organisms* including bacteria, mainly associated with the rice plant's rhizosphere, which fix small amounts of nitrogen.
4. *Heterotrophic organisms* not associated with the rhizosphere and which can fix 6– $25 \, kg \, ha^{-1}$ of nitrogen, depending on the amount of rice straw returned to the soil.

Other sources of nitrogen input include precipitation ($1–38 \, kg \, ha^{-1} yr^{-1}$); irrigation water ($6–16 \, kg \, ha^{-1}$ per crop); ground water ($0.1 \, kg \, ha^{-1} yr^{-1}$); and variable small inputs of atmospheric ammonia, organic matter and crop residue decomposition. It has been estimated that apparent nitrogen recovery in wetland rice cultivation does not exceed 40 per cent with 60 per cent in dry cultivation.

Saturated conditions also increase the availability of phosphorus and potassium. In the case of the former, nutrients conserved and more evenly distributed by control can be increased by slope terracing; by the construction of small retaining tanks or dams; by transport; and by riverine flooding. Bunded terracing of hill slopes is characteristic of highly populated areas, particularly in Monsoon Asia. This extends the area of paddy land and at the same time helps to check surface run-off from steep slopes, which can then be progressively drained down slope to augment pluvial water on the lower land. In other areas, such as Southern India, where precipitation is relatively low and the rainy season short, small temporarily bunded dams or more permanent tanks serve to collect precipitation and control its distribution. Irrigation, in the true sense of the process of transporting water from areas with an excess to those with a deficit, is not characteristic of the tropical or temperate Asian wetlands though natural flood water may be supplemented by lifting and transferring water from rivers to adjacent land.

To a lesser or greater extent pluvial water sources may be supplemented from *phreatic* (ground water) or *fluxial* sources. The dependence of paddy land on ground water alone is, however, more characteristic of the recently developed wet-rice areas of tropical Africa and Latin America where the Asiatic floodplain land forms are not found to the same extent or where, because of physical or cultural reasons, they have not been exploited for rice cultivation. In these tropical areas soil saturation, dependent on a high and seasonally fluctuating water-table, occurs in depressions or riverine swales and is the basis of 'swamp-rice' cultivation. Phreatic water is, however, not amenable to the same degree of control as that from pluvial sources. In the pluvial rice lands of Asia the existence of a high water-table helps to reduce the danger of drought during the growing season and extend the period during which the soil remains sufficiently moist for rice as well as compensating for soils with low water-holding capacity (see Fig. 11.3). In contrast, rapid deep flooding by surface run-off of lower floodplains and deltas can delay the start of the growing season until such time as the flood waters recede to a depth suitable for cultivation.

The type (variety) of rice grown, the number of rice crops taken annually and the associated cropping patterns depend fundamentally on the seasonal flooding regime (see Table 11.3) and particularly on the duration, depth and variability of flooding. The two most distinctive natural regimes are those of shallow and deep water respectively.

SHALLOW-WATER REGIMES

Shallow-water regimes are dependent mainly on pluvial sources, supplemented to a greater or lesser extent by phreatic and fluxial water. Of these, the most widespread is that in which the flood period is prolonged but variable and the growing season can be interrupted in some years by drought spells. Rarely subjected to natural flooding, rainfall is often supplemented by surface run-off from higher terraced to lower unterraced ground. Under particularly favourable conditions surface run-off can ensure continuous shallow flooding with little drought risk. Under conditions marginal to dry land, where there is a complete dependence on precipitation, the flooding regime is highly irregular and drought risk during the growing season is considerable.

Only where, as in the humid tropics of South-East Asia, the wet season is extended or irrigation water is available are two or, more rarely, three rice crops possible within the year. Throughout most of tropical and temperate Asia one crop a year is the norm either because of insufficient water in the dry season, excessive percolation under less than optimum soil conditions, low winter temperatures as in Northern China, Korea and Japan or because of a need by largely subsistence farmers to produce complementary food or cash crops. The majority of shallow-water regimes are associated with a two-phase rotation of summer paddy rice alternating with the dryland cultivation of temperate winter cereals such as wheat or barley, crop legumes (especially peas and beans), vegetables, cotton or tobacco; or under particularly favourable conditions a three-phase rotation such as cotton/winter wheat (February to October), maize/peas (May to August) and rice/cloves (June to October) as is found in Egypt.

FLOOD REGIMES

Flood regimes are dependent mainly on fluxial water, which varies in duration and depth. The shallower regimes are those where drainage and/or recession of flood water are rapid enough to prevent long periods when the depth of water exceeds 50 cm. Those with deep seasonal flooding of 1–2 m to over 5 m in depth are typical of lower river reaches and, more particularly, the middle and frequently depressed areas of broad deltas. Land can neither be bunded nor irrigated and varieties of deep-water rice (*floating rice*) which require flooding to a depth of over a metre for

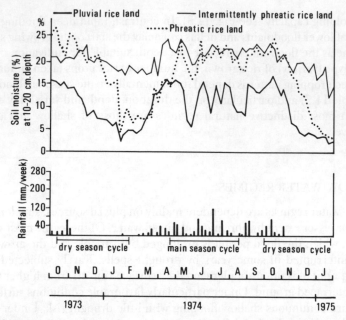

Fig. 11.3 Soil moisture content in the root zone of pluvial and phreatic rice land at the International Institute of Tropical Agriculture, Ibadan, Nigeria (from Moormann and van Breeman, 1978)

part of their growing period are cultivated. In contrast to shallow-water cultivation, the rice is planted by direct seeding on dry ground and seedling growth commences before flooding. The deep-water rice varieties adapted to this habitat elongate rapidly (*c.* 12 cm day^{-1}) as the water level rises. It is estimated that deep-water rice accounts for about one-sixth of the world's rice-growing land. The duration of deep floods frequently limits cultivation to an annual one-crop rice system. Another less extensive and less exploited flooding regime is that associated with coastal areas where flooding is dependent on diurnal or semi-diurnal tides. Known in Africa as *mangrove rice land*, rice can be cultivated only when the off-land flow of fresh water is sufficient to lower salinity to levels tolerable for rice. In some such coastal swamps sulphur levels are very high (72.2 per cent). If drained, oxidization can result in acidity of *c.* pH 2.5 at which iron and aluminium can also attain toxic levels.

The use of *irrigation* alone to flood and maintain soil moisture levels is virtually confined to the more recently developed intensive rice-growing systems in the subtropical/warm-temperate climates of Mediterranean Europe, the lower Mississippi Valley, California and scattered localities in South America and Australia. Soils with low percolation rates and high water-holding capacities are optimal for irrigated as for paddy land. In the Sacramento Valley (California) irrigated rice is grown on shallow hard-pan soils of limited value for other crops. However, not only is production highly mechanized but the modern irrigated systems differ from the traditional Asiatic paddy lands in the very high inputs of fertilizer, herbicides and pesticides; the use of new HYVs of rice; and controlled draw-off of water at intervals during the growing season and just before harvesting.

DRYLAND (OR 'UPLAND') RICE

About a sixth of the world's rice-producing land is not flooded and the crop is cultivated in the same way as other cereals (de Datta, 1975). Although dryland rice requires less water than paddy for successful cultivation, it is more susceptible to drought and needs adequate precipitation for 3–4 months of its growing period. Maintenance of the soil nutrient status is more difficult than in wetlands, with aerobic conditions giving rapid breakdown of organic matter, higher nitrogen losses and less available phosphorus. In addition, weed infestation, particularly by *Imperata cylindrica*, constitutes a serious limiting factor. Dryland cultivation of rice is dispersed throughout the humid Tropics, where it is an important element in rotations associated with shifting cultivation and bush fallowing; and in Japan, where it may receive supplemental irrigation during its later growth stages and where it tends to receive a heavy input of organic (manure) and inorganic fertilizers.

PRODUCTION AND YIELD

Yield of dryland rice varieties is in general less than that from wetland; adaptation has been to lower fertility sites, and less selection and breeding for HYVs has been carried out than in the wetland varieties. It is also more susceptible to competition from weeds and pests than paddy.

Table 11.4 Variation in rice yield with latitude

Latitude (°N/S)	Average yield (kg ha^{-1})	Approx. percentage total world area
0–10	1420	11
11–20	1550	25
21–30	1400	57
>30	3000	7

(from Grist, 1975)

Given the range of environmental conditions and methods of cultivation under which rice is grown the variation in yield from one locality to another (from 700 kg ha^{-1} in some tropical areas to 7000 kg ha^{-1} in Australia) should occasion no surprise. The higher yield in temperate as opposed to tropical regions (see Table 11.4) has frequently been attributed to climatic differences, particularly to summer temperatures nearer the optimum for growth and production; cooler nights during the growing season, which depress respiration rates; and higher sunshine intensities and longer day-length during the growing season. On the other hand, higher yields in temperate areas may be as much if not more a function of the intensive methods of production in those areas into which rice cultivation has only recently been intro-

Table 11.5 Change in nutrient balance as a result of intensive rice cropping over a 5-year period: averages of three locations and three varieties

Cropping system and fertilizer application	Total grain yield in 5 years (kg ha^{-1})	Nutrients added (kg ha^{-1})			Nutrients removed (kg ha^{-1})			Balance (kg ha^{-1})		
		N	P	K	N	P	K	N	P	K
Traditional culture:										
one crop										
No fertilizer	7 500	–	–	–	126	281	164	–126	–281	–164
HYV culture:										
double cropping										
No fertilizer	32 400	–	–	–	545	122	710	–545	–122	–710
105 kg N ha^{-1} per crop	50 100	1050	–	–	840	190	1100	+210	–190	–1100
105 kg N + 26 kg P ha^{-1} per crop	54 700	1050	262	–	920	208	1199	+130	+54	–1200
105 kg N + 26 kg P + 50 kg K ha^{-1} per crop	59 300	1050	262	262	995	225	498	+55	+37	–801

HYV = high-yielding variety.

(from Norman *et al.*, 1984)

duced (see Table 11.5). These include, because of management, the absence of flood and drought hazards; high inputs of nutrients to which *japonica* varieties give a better response than *indica* varieties; the more effective control of weeds and pests; the use of better seed; and, not least, the use of more recently developed HYVs.

While yields vary considerably, they have tended to remain stable or to show a slight decline within particular countries and discrete producing areas until the early 1960s. In 1962 the newly established IRRI produced a successful cross between a semi-dwarf *indica* variety (Taiwan) and *peta*, a traditional tall variety widely grown in the Philippines. This produced the famous new 1R8 rice, the first of the new HYVs. Shorter than most traditional varieties (*c.* 100 cm tall) with good tillering ability and a high grain to straw ratio, it was able to resist lodging. Unlike most traditional varieties it responded well to nitrogen fertilizers, giving much higher grain yield within a growing period of 125–135 days. It heralded what became known as the 'Green Revolution' (see pp. 106–7), which it was hoped would serve to keep rice production abreast of increasing population in Asia.

Although the new varieties now occupy approximately 20 per cent of the world's rice land, and rice production in India has doubled since 1975, the HYVs have created two problems. First, the early developed varieties were both pest- and disease-prone. This stimulated the production of more resistant varieties. However, one of the now most serious pests, the broom-plant hopper (*Nilaparvata lugens*), which started to attain pest proportions only in 1977 and has since then spread from the Philippines to the rest of Asia, has proved particularly virulent. It comprises several biotypes and is a virus carrier. The new semi-dwarf and leafier rice varieties, combined with year-round cropping, have created an ideal new habitat for this insect, many of whose natural predators may well have been eliminated by the early use of non-selective insecticides. In addition, with more intensive cropping, the concomitant increase in undrained ditches have provided routes along which the hopper can migrate. As yet, there is no HYV immune from attack. While the search for ever-more disease- and pest-resistant varieties goes on, new rice varieties adapted to a wide range of environmental conditions (such as varying water depth) with increased tolerance of flooding, soil deficiencies and toxicities, and with a higher protein content are continually being developed.

The second problem associated with the introduction of the HYVs of rice has been socio-economic rather than biophysical. The high yield potential of the HYVs can only be realized by a proportionately high input of fertilizers and an assured supply of water, pesticides and herbicides, at costs which the average small-scale paddy cultivator in Asia is unable to meet. The Green Revolution has, as a result, had its impact in areas where the need for subsistence food production is least but where the intensity of rice growing was already higher – where there were larger more economic holdings and/or where there was an assured supply of irrigation water and fertilizers.

Chapter 12.

Dryland agriculture

Dryland farming or crop cultivation with limited marginal precipitation is character-istic of and most extensively practised in the *semi-arid* areas of the world (Hall *et al*, 1979). While semi-aridity implies conditions neither totally wet nor completely dry, the agricultural significance of the concept can only be expressed in terms of the sufficiency of available precipitation for cultivation. The semi-arid areas lie between the deserts, where the production of rain-fed crops is not possible during the majority of years despite efforts to conserve and maximize available moisture, and the humid areas, where drought does not substantially limit crop production in the majority of years. As Thornthwaite (1948), Bailey (1979) (see Fig. 12.1) and others have demonstrated, the annual precipitation amount defining the humid–arid boundary depends on those atmospheric conditions, particularly temperature, which determine evapotranspiration rates; and a rainfall amount sufficient to produce a crop in a cool climate may be inadequate in a warmer one.

In other respects the semi-arid areas of the world have many climatic characteris-tics in common with arid areas, such as: high insolation and potential evaporation; low variable precipitation; large diurnal temperature ranges with high daytime insolation alternating with high outgoing radiation at night consequent on clear skies and low cloudiness; and strong desiccating winds. The semi-arid areas are, how-ever, distinguished from both the arid and humid areas by reason of the *variability* of the rainfall regime. The absolute variability is less than in the true desert regions, where it is so high that rainfall incidence tends to be 'episodic' rather than seasonal. A smaller variation of mean annual rainfall in the semi-arid areas can tip the balance between there being a sufficiency and an insufficiency of moisture for cultivation. Semi-aridity, then, is characterized by an alternation of periods of drought of varying length when the climate is agriculturally arid and periods when the climate is agriculturally humid. However, climatic variability can strike deeper even during years when precipitation is above average. Although more markedly seasonal than in arid areas, precipitation is variable in both time and place of occurrence; and lack of available water at the critical period during the crop growing period may, irrespec-

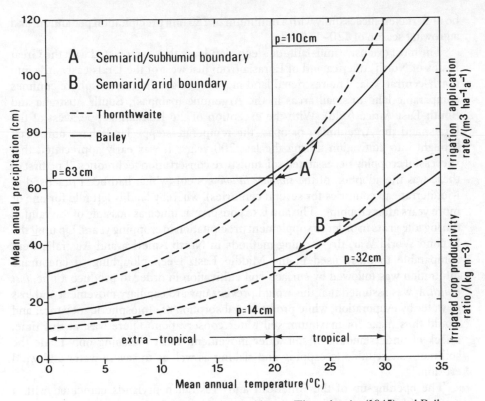

Fig. 12.1 Semi-arid/subhumid boundary according to Thornthwaite (1945) and Bailey (1978) (from Bailey, 1979)

tive of the total annual amount received, result in a drastic reduction of yield or in crop failure.

While low precipitation and a short growing season put a severe limit on agricultural options, variability in amount and incidence of available moisture make the semi-arid areas of the world those of exceptionally high-risk agriculture. A variety of farming systems have developed aimed at minimizing the risks of crop production. Common to all is the practice of restricting the number of consecutive annual crops taken from the same land. The main contrasts in cropping techniques, however, are between dryland farming in the temperate semi-arid steppe lands and that in the tropical semi-arid savanna lands – with contrasts in climate, types of crop and farming systems reinforced by differences in rural population density and in the stage of agricultural development.

The temperate mid-latitude steppe lands are characterized by a marked warm and cold season with the result that the growing season can be curtailed by unseasonable frost. Winters, particularly in higher latitudes, can be extremely variable with very severe winters, exacerbated by prolonged frost, snowfall and blizzards, alternating with milder more open conditions. Available soil moisture is usually supplemented by snow-melt in spring; and cropping, though marginal, can

197

be undertaken successfully with a minimum of 250 mm precipitation per annum and a growing season of 120–140 days.

The most extensive mid-latitude steppe lands are those associated with the Great Plains of North America and of Eurasia from just west of the Urals eastwards into north-central Asia. The restricted land mass of the southern hemisphere confines comparable land to small areas in the Argentine (pampas), South Australia and South-East Africa (veld). With the exception of the Russian steppe, west of the Urals, and the Argentinian pampas, the temperate steppe lands have only been brought into cultivation during the last 200 years. It was early appreciated that successful cropping necessitated soil moisture conservation techniques. The first of these was the adoption of the *summer fallow* system (which had been practised in Mediterranean countries for several centuries), whereby land is left idle for one or more years after cropping. The aim is to conserve as much as possible of the rainfall during idle years in order to supplement precipitation in cropping years. Up until the Second World War, dry-farming methods in North America and Australia were comparable to those used in the Middle East. Deep ploughing to encourage infiltration was followed by surface-soil cultivation in order to produce a fine *dust mulch*. It was assumed that this would prevent upward capillary movement and loss of water by evaporation, while promoting absorption of atmospheric moisture, and would thus make for maximum soil water conservation. There was, at this time, a lack of understanding of soil water movement. Dust mulching only made the soil more susceptible to capping and slaking as well as to severe water and wind erosion.

The opening-up of the American and Australian drylands coincided with a period of high demand for food at the beginning of the twentieth century, when rainfall was above average (Ridley *et al*, 1980). With the onset of the disastrous droughts of the 1930s (the 'dirty thirties') these fine mulch soils suffered severe wind erosion with the creation of the notorious Dust Bowl conditions, particularly in the southern Great Plains, and dust storms which affected the atmosphere as far as the east coast of the USA. The resulting encouragement and development of soil conservation methods in the 1930s and early 1940s brought about a modification of the practice of summer fallowing (Stobbs, 1979). Shallow ploughing to eliminate weeds and conserve crop residues on the surface (i.e. *stubble mulches*) was adopted and was facilitated by the use of specially adapted tillage implements such as the wide-blade cultivator, the chisel plough, the field cultivator and the rod-weeder. No tillage was undertaken from the time the last crop was harvested; three tillages were normally undertaken in early June – and more than three in wetter years – in order to control virulent perennial weeds such as couch or quake grass (*Agropyron repens*).

Summer fallow is particularly important for weed reduction because, with each tillage operation, seeds are exposed to ideal germination conditions. Some subhumid soils need up to ten cultivations since, if the soil is moist for long periods, weed seeds will set and hence be able to re-infest. The seeds of wild oats (*Avena fatua*), in particular, have a long dormancy which requires cool moist conditions for it to be broken. However, since the end of the Second World War, the introduction of herbicides have made weed control possible without the necessity of cultivating the summer fallow. This reduces the risk of soil erosion and nitrogen loss but tends to aggravate the salinity problems which are exacerbated by summer fallowing and increased soil water evaporation.

The amount of water actually conserved by these means is small: somewhere between 16 and 20 per cent of the total annual rainfall. As Russell (1977) notes, only water in the zone over 20 cm from the surface can be stored since most of that in the immediate subsurface is lost by evaporation. Measurements in the northern prairies of the USA revealed that of 57.7 cm precipitation received during the 21-month fallow period, 10–19 cm were retained; and 6.5 cm of this were conserved during the first 9 months. Furthermore, fields that remained untilled during the 9 months from harvest to seeding conserved only 1–2 cm more than those tilled. At the same time other soil conservation techniques were introduced including (Skidmore *et al*, 1979): contour ploughing; strip cultivation with an alternation of cultivated and fallow land aligned across the direction of the prevailing wind; a temporary cover crop such as a fast-growing millet; or a stubble mulch with a cloddy compacted soil surface.

In addition, increasing emphasis was given to the role of the summer fallow as a means of regenerating soil nitrogen as well as conserving moisture. Despite the relatively high nutrient status of the grass-based incompletely leached soils of the steppe lands, tillage promoted rapid loss of organic matter – the main nitrogen store. Nitrogen levels have continued to decline, with many soils in the USA now containing less than 50 per cent of their original store. Not only were some cultivated areas, particularly on the drier margins, put back to permanent grazings but the establishment of a sown 'grass-break' instead of the idle fallow helped to stabilize soils by accumulating organic matter and encouraging the development of a crumb structure. A more recent revolutionary change in dry-farming methods in North America has been the change from cropping with summer fallow to zero tillage. This was initiated by the introduction, in 1961, of *paraquat* – a broad-spectrum herbicide that leaves no active soil residues. Crop yields under zero tillage have been shown to be comparable, if not better, than under conventional tillage methods. The stubble and trash left on the surface retains twice as much snow, which also melts more slowly, than on the ordinary fallow. Other advantages include a reduction of the soil salinity and of annual weeds, with a greater conservation of organic matter. On the other hand, perennial weeds can become more difficult to control. Fertilizer placement is less easy and soil temperatures at the time of seeding are lower than in tilled land. While zero tillage saves time and cost of production, its widespread adoption is determined more by high demand coupled with lower crop prices than by its intrinsic potential for soil conservation.

Even with the most efficient system of soil water conservation, cropping options in these semi-arid areas are limited, not only by low rainfall but also by a short growing season. Temperate cereals – wheat, barley and rye in particular – are well adapted to climatic conditions in which available moisture is concentrated in a short period and where high amounts of sunshine and high temperatures promote rapid growth and early maturation. The opening up of the temperate semi-arid lands for large-scale wheat production in the late nineteenth century came as a result of growing demand in both North America and Europe, the introduction of fast cheap transport and the development of agricultural techniques suited to the environmental conditions. In the last respect effective ploughing had to await the introduction of specialized ploughs: the steel chisel plough, which was capable of breaking the tough grassland sod in the North American prairies and the jump plough, adapted for use in the cut-over eucalyptus woodland of Southern Australia. In addition, new flour-milling techniques enabled the potential of drought-resistant hard wheat varieties for bread-making to be fully realized.

In the semi-arid areas where severe winter conditions and frozen ground curtail the cultivation period, rapidly maturing spring-wheat varieties are grown; less severe conditions allow the cultivation of autumn-sown winter wheat, varieties of which tend to need low winter 'vernalizing' temperatures before they can restart growth and mature quickly in the spring. Yields of winter wheat are normally higher than those of wheat sown in spring. Yields of both, however, are limited by low precipitation and are inevitably lower than for varieties grown under humid climatic conditions. Winter-wheat yield in Western Europe is some three times that in semi-arid North America and Australia, and six times that in the steppe lands of the USSR. However, large holdings compensate for low yields. A high level of mechanization makes for the speed and timeliness of operations essential to minimize the hazards of seasonal variability and to give high outputs relative to man-hours in areas where farm population is sparse.

Wheat is still the dominant crop of the temperate dry-farming areas, which produce 90 per cent of the world's bread flour. The limits of cultivation, however, have advanced and retreated with the alternation of dry and wet periods. Since the disastrous droughts of the 1930s some very marginal land, formerly cultivated, has reverted to grazing; in contrast, under more favourable rainfall conditions or where it is economic to tap ground water for irrigation, the former monoculture of wheat has been modified and cropping has become more diversified. Ley farming, involving the rotation of small grains (wheat, barley or rye) with short-season annual legumes (alfalfa, lupins or clover) which can reseed themselves, and the cultivation of alfalfa and sugar beet by irrigation have allowed the integration of livestock rearing with crop production. This integration of extensive and intensive cropping with livestock has developed furthest in the USA, where with a ratio of 100–200 ha of irrigated fodder to 800–1200 ha of summer-fallow dry-farmed cropland, over a hundred cattle can be reared and/or fattened. Livestock, however, still play a relatively minor part in the dry-farming areas except in South Australia, where the introduction of subterranean clover (*Trifolium subterraneum*) in the 1930s allowed the integration of wheat and sheep. Introduced to Australia and New Zealand from Mediterranean Europe, subterranean clover is best suited to areas with warm summers and mild winters, with 500–600 mm of winter rain and a growing season of 5–9 months (Langer and Hill, 1982). It is both persistent and hardy. It can survive under severe summer drought because of its habit of burying its seed just before the start of the dry season; and its large hard seeds can remain viable for long periods. Prostrate growth allows the plant to withstand close grazing. In contrast, in the Argentine, cattle ranching and wheat cultivation on the basis of shifting cultivation were associated with the early development of *estancia* farming.

MEDITERRANEAN AGRICULTURE

The land area peripheral to the Mediterranean Sea is one of the most important 'cradles of agriculture' in the world. Sedentary agriculture based on seed planting developed early – over 5000 years ago. Many of the progenitors of the most important food and feed crops originated and were disseminated from this area. Situated at the 'crossroads' between Europe, Africa and Asia, exotic species from east and west, from temperate and tropical climates, were brought together in the traditional farming systems.

The Mediterranean environment in Europe, California, South Africa and South-West Australia is characterized by a warm-temperate semi-arid climate. This differs from the semi-aridity of the steppe and savanna lands in that the rainy season coincides with the winter period when insolation and temperatures are at their minimum. There is an alternation of a humid temperature growing season in the winter with an arid tropical summer, when precipitation is low, potential evapotranspiration is high, and there is an absolute soil water deficit. In addition, the Mediterranean areas coincide with recently folded and faulted sedimentary rocks and a correspondingly broken upland to mountainous terrain. The contrast between closely juxtaposed coastal and river-basin lowlands and steep mountain ridges is sharp. Rapid run-off of snow and ice-melt in spring makes for high flood risks. The intensity of terracing, which has left its mark on hill slopes up to altitudes of 700–800 m in Mediterranean Europe, was a means not only of extending the area of cultivable land but of checking run-off and soil erosion.

Three types of land-use characterized traditional Mediterranean agriculture and they still persist, though their crop components and their relative importance has changed, particularly since the 1940s. One was the cultivation of temperate crops during the winter growing season, with an emphasis on autumn-sown (winter) wheat and barley. The former was the hard 'durum' wheat with low rainfall requirements, deficient in gluten and more suited to pasta than to bread-making. The cereal crop was alternated with a summer fallow on which sheep grazed the stubble and aftergrowth.

The second type of land-use was the cultivation of subtropical, frost-sensitive and drought-resistant trees and shrubs such as olive, grapevine and almonds. Irrigation allowed the climatic potential of high summer temperatures and insolation to be exploited and thereby to give an all-year growing season. Both humid-tropical crops such as citrus fruit, melons, pineapples, bananas and warm-temperate fruits and vegetables could be grown in areas where suitable soils and level terrain coincided with relatively easy access to a reliable source of water. The Moors, who introduced irrigation techniques to the Mediterranean, gave the Spanish term *heurta* (or garden) to these diverse and very fertile 'summer-green' lands.

The third type of land-use was rough-grazing. As in other semi-arid regions cropping and livestock tended to be segregated, and the cow was subordinated to the sheep and the goat as milk and meat producer. Both the sheep and, more particularly, the goat could maintain themselves and make more efficient use of the uncultivable land of the garrigue and maquis of the Mediterranean. These are dominated by hard-leaved evergreen shrubs better adapted to summer drought than the grasses and other herbaceous species which are the usual diet of the cow.

The traditional agriculture has left its imprint on the landscape. Increasing urbanization, expanding markets for what, before the 1940s, were luxury foods (particularly tropical fruits), intensification of agriculture and financial subsidies have effected increasingly rapid changes in farming systems. Much of the poor soil and upland areas, formerly terraced and/or grazed, have been abandoned. The winter cereal crop has declined and has been replaced by fast-growing high-yielding hybrid ryegrass, which provides a more nutritious feed for sheep, whose numbers, as a result of EEC subsidies and demand, have increased. The unirrigated tree and shrub crops have persisted but have become more specialized and concentrated in areas where high production of fruits can be obtained by means of increasing inputs and intensification of agriculture.

Irrigated agriculture has become relatively more important in extent and productivity; and intensive irrigated agriculture is now the dominant agro-ecosystem in the Mediterranean environment of the Old and New Worlds. Modern irrigation systems are combined with high inputs, particularly of pesticides and of energy in the form of heaters and wind-fans, to minimize the risk of air frost. Citrus and other fruit growing has become increasingly specialized with an ever-greater diversity of 'varieties' which ensure all-year-round production and product specialization. The constraint of a long maturation period, so characteristic of tree crops, has been mitigated by the production of new varieties by grafting. In addition, irrigation has permitted the production of 'early' fruit and vegetables – strawberries, peaches, lettuce, potatoes etc. – that can compete successfully with those grown in higher cooler latitudes near to the markets of both Europe and North America. Further, with irrigation, double and triple cropping of some crops is possible. Two potato crops are now common, while the cultivation of feed crops such as lucerne (which can give up to five cuts in a season), ryegrass and field beans has become profitable with the development of dry-lot dairy farms located near to the now-large urban markets in the Mediterranean areas of the world. While irrigation may have overcome the constraints and released the full potential of the Mediterranean climate in certain areas, it has created another set of environmental constraints which are as, if not more, difficult to overcome than the original ones (see Chapter 13).

TROPICAL DRYLAND FARMING

The most extensive semi-arid areas in the Tropics and sub-Tropics occur in Africa, particularly in the Sahel zone which cuts across North Africa between the savanna and the Sahara Desert, and in the karoo of South Africa between the Kalahari Desert and the veld; and in North and East Australia; with smaller areas in the Indian subcontinent and North-East Brazil. Climatic conditions are, from an agricultural point of view, comparable to those of temperate semi-arid regions. Temperatures, however, are higher all the year round and hence the precipitation limits for successful crop cultivation are also higher; and there is no cold winter season as in temperate latitudes.

Soils, however, are generally poorer. In contrast to the temperate grassland soils, with younger, less-weathered and largely sedimentary parent material, soils in the semi-arid Tropics are mainly derived from metamorphic and igneous rocks which have been subjected to a very long period of weathering and leaching. In addition, they are low in organic matter and nutrients – particularly nitrogen and phosphorus – and iron-indurated horizons at or near the soil surface are common. In the Tropics, poor soil fertility increases the restrictions already imposed by a low and variable precipitation on successful agriculture and crop yields. The rotation of cropping with fallow periods of varying length is, as in the humid Tropics, a means of maintaining soil nutrient status and crop yield rather than of just conserving soil moisture. However, as in temperate regions, low rainfall restricts the type of crops that can be grown without irrigation (Gibbon and Pain, 1985). Subsistence staples are mixtures of the more drought-resistant sorghums and millets. Varieties of one or both these cereals are selected to ensure maximum production in the face of a wet season of variable and unpredictable duration. With increasing aridity, millet tends to displace sorghum, which is more demanding of soil water. In addition, the latter, in comparison to millet, produces a bulkier residue which is slow to decompose during the dry season and hence 'locks up' too much nitrogen in its residues.

In comparison to the humid Tropics, the variety of crops grown is less: cow peas and groundnuts, both adapted to the short growing season, are valuable sources of protein, while cassava tubers, tolerant of poor soil conditions and capable of remaining in the ground without deterioration during a long dry season, give high yields of carbohydrates. Cropping frequency is also curtailed by climatic conditions. Two crops are possible only under exceptional soil water conditions, such as occur on the black-clay regur soils of the Indian Deccan or where water conservation in tanks or small reservoirs is practised. The most widespread system is a one-year crop followed by one or more years of bush or grass fallow. When recultivated, the fallow ground is normally burned just before the onset of the rainy season, and the crop seed is sown in stages after the commencement of the rains. Areas of permanent and semi-permanent agriculture are (as in humid Africa) confined to 'kitchen' gardens and enclosed 'compound land' or to small favourable sites, e.g. vlei soils outside the compound.

Low and very seasonal precipitation, as in the temperate drylands, has tended to result in a functional separation of extensive pastoral farming based on migration between the arid and semi-arid land and sedentary cultivation where there is sufficient water available for at least one crop and for domestic purposes throughout the year (Ruthenberg, 1980). Not only do these two types of agricultural activity co-exist, particularly in semi-arid Africa, but there is a higher degree of functional integration, of ecological symbiosis, between the two systems than in temperate areas. The cultivators generally keep goats, sheep, hens and cattle if sufficient grazing is available. Crop stubble is open to grazing by both sedentary and migrating herds of cattle. The former tend to be coralled at night and grazed near the settlement to allow maximum collection and use of dung. The latter graze harvest stubble and fallows, and the land gains through addition of manure, trampling of crop residues and stripping of cereal stalks, as well as the breakdown of old cultivation ridges. In many areas of semi-arid Africa, acacia trees (*Acacia albida*) are either retained or planted on cultivated land. Their pods and deciduous leaves form a valuable nutritious fodder, which increases the amount and value of livestock droppings in their vicinity.

Another important form of integration is that whereby cattle owned by cultivators are tended by pastoralists. This type of integration occurs increasingly, in situations where cultivators own more cattle than can be maintained on their land. As exemplified by the Tanzanian 'cotton cycle' it has been increasing as a result of the increased income from cash crops such as cotton, groundnuts etc. Without an alternative outlet the income is invested in cattle, the social and cultural value of which lies in their numbers rather than their quality or economic value.

However, as a result of increasing population pressure, combined with political action designed to ensure and protect rights to cultivate land within defined individual or tribal boundaries, not only has cultivated land been increasing at the expense of pasture in semi-arid Africa but the fallow period has been reduced to less than a year with drastic consequences for soil fertility. At the same time, increase in livestock numbers is exerting higher grazing pressures on the diminishing area available. While solutions to the problem of declining productivity of the land as a result of overutilization by a rapidly growing and expanding population of subsistence cultivators are largely political and technical, they are made more difficult by the unfavourable cost to benefit ratio of the necessary investment, because of the inherently low productive capacity and the high production risks imposed by the semi-arid climate.

DESERTIFICATION (see Fig. 12.2)

The expansion of arid at the expense of the surrounding semi-arid and even subhumid lands was an early noted phenomenon. However, until the Sahelian drought of 1968–72 it had elicited little more than academic interest. The United Nations Conference on Desertification in 1977 highlighted the global scale and increasing rate of this process, which was officially defined as the spread of desertlike conditions of low biological productivity into areas outside the previous desert boundaries. A large-scale international research programme was initiated to analyse the causes, to understand the process of desertification and to seek remedies.

It had originally been assumed that the spread of desert-like conditions was a consequence of decreasing rainfall – of increasing climatic aridity. However, there is little evidence to support the assumption of such a long-term global trend. It has now become clear that the causes of desertification are complex and multivariate and that desertification is a continuing process resulting from the misuse of land resources at times of climatic stress (i.e. of drought), particularly in the semi-arid/subhumid zones marginal to the true deserts of the world. Recent studies have confirmed that high rainfall variability and drought risk are inherent characteristics of semi-arid climates. Further, evidence from a variety of sources testifies to a long history dating from the Quarternary period of alternating periods of variable length of above

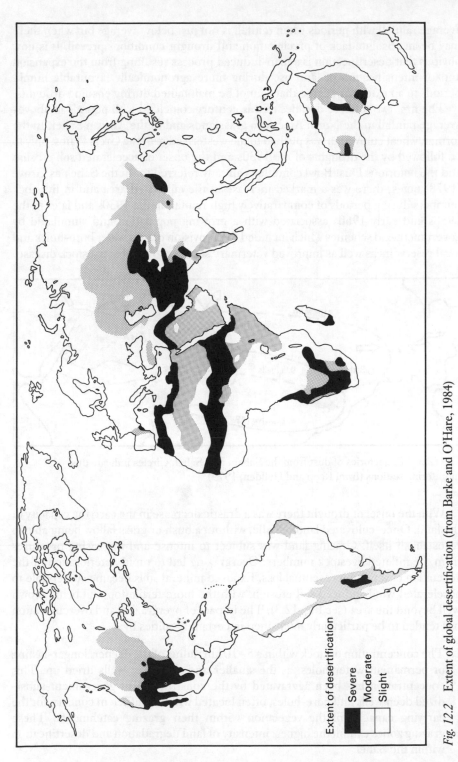

Fig. 12.2 Extent of global desertification (from Barke and O'Hare, 1984)

Extent of desertification

Severe
Moderate
Slight

average rainfall with periods when rainfall is not just below average but when there may be an absolute lack of precipitation and drought conditions prevail. It is now obvious that desertification is a man-induced process resulting from the expansion and/or intensification of land use during more agronomically favourable humid periods to an extent or degree that cannot be maintained during ensuing droughts.

The first two to three decades of this century coincided with periods of above-average rainfall in the North American grasslands and in the African Sahel. In the former, wheat cultivation was pushed to the western edge of the Great Plains, only to be followed by the droughts of the 1930s and the onset of accelerated soil erosion and the notorious Dust Bowl conditions already referred to. In the Sahel, as Grove (1978) notes, there was a marked increase in the cultivated area and in livestock numbers during periods of comparatively high rainfall in the 1930s and later in the 1950s and early 1960s associated with a growing population and stimulated by government-aid schemes which included the provision of new water bore-holes and local reservoirs as well as improved veterinary services to combat livestock disease.

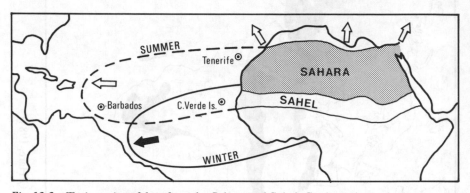

Fig. 12.3   Trajectories of dust from the Sahara and Sahel. Circles indicate dust monitoring stations (from Rapp and Hellden, 1979)

With the onset of drought there was a drastic decrease in the carrying capacity of the land. Once-cultivated land fell idle, without a bush or grass fallow being able to re-establish itself. Grazing land was subject to intense and increasing pressures from the inflated livestock numbers. Overgrazing led to rapid deterioration of the vegetation cover and its eventual loss. Exposed mineral soils became susceptible to accelerated wind-induced soil erosion, with the more fertile top-soil being blown well beyond the area (see Fig. 12.3). The impact of overgrazing and overcultivation has tended to be particularly intensified in certain localities by:

1. The concentration of stock within a 5–10 km radius of the deeper, longer-lasting or permanent water-holes, as the smaller and shallower wells dried up. This concentration has been aggravated by the construction of government-subsidized deep wells and bore-holes, often located without sufficient concern for the carrying capacity of the vegetation within their grazing catchments. These grazing zones exhibit the highest intensity of land degradation and desertification within the Sahel.

2. Decrease in the length of fallow periods from over 20–30 years to as little as 3 years or less in three or four decades.
3. Increased demand for fuelwood, particularly for domestic purposes, which has accelerated the process of deforestation and, consequently, of soil erosion.

As a result of these more localized pressures, the rate and intensity of desertification is variable within the semi-arid zone, being initiated at focal points which grow outwards and eventually coalesce (Rapp, 1974; Rapp and Helleden, 1979).

It has also been suggested that desertification may have a positive climatic feedback. Increase in surface albedo, as a result of the exposure of bare mineral soil, may effect an increase of heat loss by outgoing radiation; and this could reinforce the atmospheric subsidence characteristic of the Intertropical Convergence Zone, thereby further inhibiting precipitation and increasing climatic aridity. The strength of this hypothesis, however, depends on the extent to which the increase in atmospheric dust may counteract that in surface albedo.

Overcultivation and overgrazing in semi-arid areas such as the Sahel also have had a deleterious impact on the hydrological cycle. Exposed soil surfaces become susceptible to sun-baking, soil-sealing by torrential rain storms and to compaction by animal and human trampling. Infiltration capacity is reduced, and water loss by surface evaporation and surface run-off is increased with a decline in available precipitation. As soil water reserves and soil water recharge decline, soil aridity with declining productivity results.

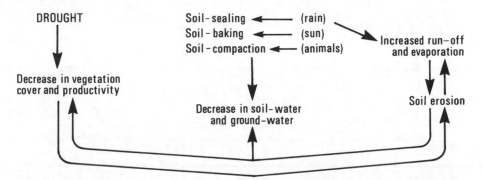

Fig. 12.4 Cycle of soil degradation as a result of prolonged drought and overgrazing

Man-induced (i.e. cultural) desertification is continuing, despite the application of soil conservation techniques in temperate drylands. However, nowhere is it expanding at such a rate and with such environmental and socio-economic repercussions as in Africa. In 1975 it was estimated that in the Sudan the desert land had expanded south 90–100 km in *c.* 17 years, while the influence of dust storms generated in the Sahel now extends well beyond the area. Today it is realized that desertification can become a dynamic self-accelerating process resulting from positive feedback mechanisms which create a downward spiral of land degradation (Fig. 12.4). After each recurring drought and the onset of the rains the recovery of vegetation and of agricultural productivity becomes increasingly difficult, if not impossible, as the pressures of livestock and people return to, or more commonly exceed, previous levels. The concern now is that in the Sahel, in particular, the process of desertification has attained a threshold beyond which biological recovery, if not impossible, is difficult because of the scale of the area and the associated socio-economic and political constraints.

Chapter

13

Irrigation agriculture

In many parts of the world there is a deficiency of soil water for crop production for part or all of the year and arable agriculture is dependent to a greater or lesser degree on irrigation. In semi-arid regimes irrigation is also used to extend cultivation, increase yields and to ensure against high rainfall variability. In humid areas supplemental irrigation is used to overcome soil water deficits that may occur at critical periods during crop growth. In Britain a soil-water drought has been defined as a deficit of over 150 mm during the period April to December (MAFF, 1982b).

Irrigation, in the strict sense, has been defined as the transport of water from a source of supply to land where there is a lack or deficit for agricultural purposes; and desert land can be defined as that on which crop production is not possible without irrigation. The warm deserts occupy about a sixth of the earth's surface. These are areas where, because of clear cloudless skies and high insolation rates, potential annual photosynthesis (Chang, 1970) is at a maximum. Also, the annual tempera-ture regime is such as to allow a year-long thermal growing season. Potential agricultural productivity, given an assured supply of water, is considerable as is demonstrated in areas formerly and currently irrigated; and yields are much higher than those from dry rain-fed land under similar climate and soil regimes. In addition, they are very stable provided soil conditions do not deteriorate. In Fig. 13.1 Stanhill (1986) contrasts irrigated wheat yields in Egypt with non-irrigated yields in Israel. Those in the former are on average 50 per cent higher than those in Israel; and they have a standard deviation of 7 per cent, which is comparable to that in England, in contrast to 26 per cent in Israel. Productivity of irrigated land is further increased by perennial cropping and Egypt currently produces some five crops every 3 years from each cultivated hectare.

The amount of cultivated land completely dependent on irrigation occupies less than 10 per cent of the total, while that potentially irrigable is estimated at no more than 20 per cent. Irrigable land is, in fact, severely limited by the combined constraints of land forms, soil, and water supply. Successful economically viable irrigation needs a level graded land surface and sufficient depth of permeable soil to

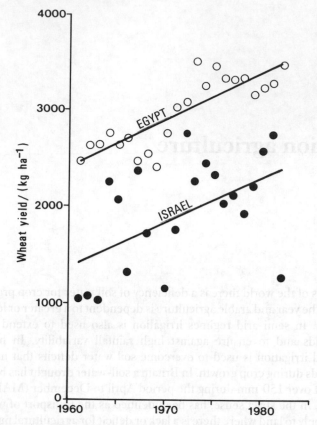

Fig. 13.1 Size and stability of wheat field yield in irrigated and non-irrigated arid-zone agriculture. Lines = linear regression of average national yield, Kg ha^{-1} (y) on year of harvest (x). Open circles: Egypt (irrigated land) F = 50; Black circles: Israel (non-irrigated land) F = 7 (from Stanhill, 1986)

allow a water-infiltration rate and a water-holding capacity such as to ensure the maintenance of sufficient soil water for successful crop production. In both arid and semi-arid areas these conditions are satisfied particularly by the deep alluvial deposits associated with coastal plains, river floodplains, some inland drainage basins and the extensive coalescing hill-foot fans (*bahajas*) which are a characteristic feature of arid and semi-arid areas. Because of their naturally sandy texture, into which water can infiltrate more rapidly than elsewhere, such deposits are also important water catchments. Optimum infiltration for irrigation needs are provided by fine sandy to sandy loams. On coarser material the rate will be too great relative to the water-retentive capacity of the soil; on the finer clays infiltration rates may be so slow as to engender high evaporative loss of water and to limit the depth to which available water can penetrate.

Irrigation is basically dependent on an assured supply of easily and economically accessible water of suitable quality. The two main sources of supply are *surface water* and *ground water*. *Surface water* is limited in arid and semi-arid areas. Rainfall is insufficient to maintain constant river flow as a result of high evaporation, rapid surface run-off and high infiltration rates. Flash-floods in otherwise dry wadis or arroyos follow intense downpours in deserts. Rivers tend to be ephemeral with intermittent flow. The most important surface-water supplies are provided by exotic rivers whose catchment lie in high rainfall areas at high altitudes beyond the arid lands. Water may be drawn direct from rivers or impounded in reservoirs which are usually designed to ensure 1 or even 2 years supply. Water loss by evaporation is inevitably high from such open-water surfaces and long-distance transport is usually effected by pipeline. Withdrawal of surface water is, however, limited by the necessity, usually required by law, to ensure that the natural flow of the river is maintained at an acceptable level.

Ground-water resources are considerable. Their use is, however, determined by their accessibility to irrigable soils in terms of depth and location. Withdrawal of water inevitably proceeds at a rate faster than recharge even by deliberate 'watering'. As the level of the water-table drops, so the energy and cost of pumping from ever greater depths increases. Fall in water level due to extraction in one locality often results in a draw-down in another beyond the immediately exploited area. Further, in coastal regions, such as Southern California, pumping has resulted in a lowering of the water-table below sea level with the consequent incursion of brackish or salt water. Subsidence of land surfaces is another hazard associated with a falling water-table.

No water used for irrigation is 'pure' in the literal sense of the word. Both surface and, to an even greater extent, ground water in arid and semi-arid areas usually contain a higher load of soluble and suspended materials than that elsewhere. The higher chemical content is a result of high evaporation rates; and of relatively little-weathered rocks and soil from which minerals have not been leached as in humid areas. Mineralization of water may also be the result of fossil brines which have been trapped underground or, as has been demonstrated in Israel, of the evaporation of air-borne salt spray.

In arid areas rainfall is only sufficient to wet the soil to a limited depth, which becomes shallower with increasing aridity. Under extreme conditions the only perennial plants, other than succulents, than can survive are the *phraeophytes*, whose roots are long enough to tap a water-table which may be several metres below the surface. In the absence of leaching, products of weathering accumulate in the upper part of the soil profile. These include calcium carbonate ($CaCO_3$) and a variety of soluble salts (see Table 13.1). The latter distinguish the arid from the subhumid, incompletely leached, grassland soils in which calcium-enriched horizon is present but in which the soluble and therefore easily leached salts do not accumulate.

In the aridisols, calcium and magnesium are usually present as insoluble carbonates or as slightly soluble gypsum ($CaSO_4.2H_2O$). Calcium carbonate is always present in one form or another – as hard or soft nodules or concretions; or as cemented horizons or pans (*caliche*) of varying thickness. The soluble salts are frequently dominated by sodium, chlorine, potassium, sulphates, carbonates, bicarbonates (HCO_3)$^-$, sometimes nitrates, as well as a diversity of trace elements such as boron, lithium, manganese, iron and zinc. The result is that the arid soils are alkaline in reaction, with pH values over 7.

Table 13.1 Soluble salts in soils; the carbonates and gypsum, not classified as readily soluble salts, are included for comparison

Salt*	Chemical formula (in form found in soils)	Frequency in soils	Solubility	Toxicity to crops
Calcium carbonate	$CaCO_3$	very common	very low	nil
Calcium bicarbonate	$Ca(HCO_3)_2$	see notes	low	nil
Sodium carbonate (soda)	Na_2CO_3	uncommon	high	nil
Calcium sulphate (gypsum)	$CaSO_4.2H_2O$	common	low	nil
Magnesium sulphate	$MgSO_4.7H_2O$	very common	high	high
Sodium sulphate	Na_2SO_4	very common	high	moderate
Magnesium chloride	$MgCl_2$	uncommon	very high	high
Sodium chloride	$NaCl$	very common	high	high

* Magnesium carbonate, potassium carbonate, potassium sulphate, calcium chloride and potassium chloride are rare in arid soils. Nitrates only accumulate in exceptionally arid deserts.

(from data in Young, 1976)

THE SALINITY HAZARD

The type and yield of crops grown under irrigation in arid and semi-arid regions are determined by the degree of salinity or alkalinity of the soil environment. The presence of salts inevitably subjects the plant to direct and indirect stresses. The direct specific stresses are those associated with changes in the osmotic potential of the soil water, with unavailability of nutrients and with the presence of toxic ions; the indirect with the particular physiochemical properties of saline/alkaline soils.

Table 13.2 Crop response to soil salinity

EC soil-solution extract (mmho cm^{-1} at 25°C)*	Crop response
0–2	largely negligible
2–4	yields of sensitive crops may be restricted
4–8	yields of many crops restricted
8–16	only salt-tolerant crops
>10	very few salt-tolerant crops

EC = electrical conductivity.

(from Hills, 1966)

The most prevalent types of salinity are those caused by chlorides and sulphates of sodium and calcium. Aridisols differ from humid soils in that sodium makes up a higher proportion of the total cation exchange capacity, while hydrogen cations are present in very small amounts. The *salinity status* of a soil is, for agricultural purposes, usually expressed as the electrical conductivity (EC) in millimho per centimetre of the soil solution – the EC being proportional to both the salt concentration and to the osmotic pressure of the solution. A soil is considered to be saline if the EC is greater than 4 mmho cm^{-1} (Meiri and Levy, 1973) (see Table 13.2). Increase in the osmotic pressure of the soil solution, particularly in the upper part of the soil zone, makes the movement of water and nutrients into the plant increasingly difficult. Most crops are, to a greater or lesser degree, sensitive to salinity with the result that it is one of the most important factors determining yield in arid and semi-arid areas. Crops vary in their *salt tolerance* (see Table 13.3) and, above a certain level, growth rates and yields decrease. The salt-stressed plant can exhibit symptoms similar to those produced by drought: smaller bluish-green leaves; reduced size and a low shoot to root ratio; necrosis and leaf burn; and wilt.

Yield reduction is gradual and there is no well-defined critical salinity threshold. Crop tolerance has, therefore, been defined by an arbitrary value of yield loss expressed as a percentage of the yield expected under normal non-saline conditions: a 20 per cent level is used in the Netherlands and North Africa; 50 per cent in the USA. However, the concept of salinity tolerance is complex: tolerance levels vary in time and space according to the type of crop, its growth stage and the length of time it is exposed to high salt levels, as well as the depth and moisture content of the soil in which it is grown.

The effect of salinity on crop yield depends on the type and harvested part of the crop. Citrus fruits, almonds, avocados, grapes and many deciduous trees are extremely salt-sensitive. The vegetative growth of barley, wheat and cotton may be drastically reduced without a decline in seed yield. In contrast, rice can grow in saline conditions but will not produce grain. In all crops the number and size of fruit and of vegetable organs are reduced. Crop sensitivity also varies during growth, the young plant being much less salt-tolerant than the established crop. In some cases, however, salinity may improve crop quality, as in the case of sugar beet, melons and carrots, by effecting an increase in sugar content. In others, such as forage crops, the nutritional value is usually reduced and the salt content of the plants attains levels which can impair the health of grazing livestock, particularly cattle.

While crops respond to total soil salinity, their performance is also affected by nutrient availability and/or toxic conditions resulting from the specific effect of particular salts. Chlorine, sodium and boron are the most common causes of toxicity, with toxicity levels highest for chlorine and lowest for boron. The latter is one of the essential trace minerals or micronutrients and is present in most saline soils. It can, unfortunately, become toxic at concentrations only slightly above those required for optimum crop performance (see Table 13.4). At levels of 2–4 p.p.m. boron restricts cropping to a few more-tolerant crops such as sugar beet, alfalfa, date palm and some brassicas. Indeed the difference between the level of boron deficiency and that of toxicity is very narrow. Boron is, in addition, resistant to leaching.

Sodium, chlorine and, to a lesser extent, sulphate and bicarbonate can, at certain

Table 13.3 Tolerance of various crops to exchangeable sodium percentage (ESP)

Tolerance to ESP and range at which affected	Crop	Growth response under field conditions
Extremely sensitive (ESP = 2–10)	deciduous fruits nuts citrus fruits avocados	sodium toxicity symptoms even at low ESP values
Sensitive (ESP = 10–20)	beans	stunted growth at low ESP values even though physical condition of soil may be good
Moderately tolerant (ESP = 20–40)	clover oats tall fescue rice dallis grass	stunted growth due to both nutritional factors and adverse soil conditions
Tolerant (ESP = 40–60)	wheat cotton alfalfa barley tomatoes beets	stunted growth usually due to adverse physical condition of soil
Most tolerant (ESP > 60)	crested and fairway wheatgrass tall wheatgrass Rhodes grass	stunted growth usually due to adverse physical condition of soil

(from Yaron *et al*, 1973)

levels, produce toxic symptoms in particular crops. They can also disturb the plant/ soil nutrient balance by promoting or inhibiting the uptake of other ions. The most sensitive to disruption is the *calcium–sodium balance*. Sodium, sulphate, bicarbonate and magnesium all tend to inhibit calcium uptake, and thereby create a calcium deficiency. An increase in the proportion of sodium promotes higher levels of soil alkalinity. Soils in which 15 per cent of the total cation exchange capacity is occupied by sodium and with a pH of over 8.5 are so alkaline as to inhibit the growth of all but a very few crops. The presence of sodium carbonate gives rise to levels of alkalinity at which phosphorus, iron, manganese and zinc become unavailable in the absence of organic matter; and, because of the ease with which phosphorus can be fixed in an unavailable mineral form, aridisols are often phosphorus-deficient.

SALINIZATION AND ALKALINIZATION

While soil salinity is an inherent agricultural problem in arid and semi-arid areas of the world, the use of irrigation itself inevitably leads to increasing soil salinization

Table 13.4 Limits of boron in irrigation water for crops of different degrees of boron tolerance

Tolerant (limit: 4.0 p.p.m.; req.: 2.0 p.p.m.)*	Semi-tolerant (limit: 2.0 p.p.m.; req.: 1.0 p.p.m.)	Sensitive (limit: 1.0 p.p.m.; req.: 0.3 p.p.m.)
athel (*Tamarix aphylla*)	sunflower (native)	pecan
asparagus	potato	walnut (black and Persian, or English)
palm (*Phoenix canariensis*)	cotton (Acala and Pima)	Jerusalem artichoke
date palm (*P. dactylifera*)	tomato	navy bean
sugar beet	sweetpea	American elm
mangel	radish	plum
garden beet	field pea	pear
alfalfa	ragged-robin rose	apple
gladiolus	olive	grape (Sultanina and Malaga)
broad bean	barley	kadota fig
onion	wheat	persimmon
turnip	corn	cherry
cabbage	milo	peach
lettuce	oat	apricot
carrot	zinnia	thornless blackberry
	pumpkin	orange
	bell pepper	avocado
	sweet potato	grapefruit
	lima bean	lemon

* Limit = concentration at which boron becomes toxic; req. = concentration required for optimum performance; p.p.m. = parts million–parts^{-1}.

(from Yaron *et al*, 1973)

and alkalinization. Several processes are involved. Under conditions of high evaporation the depth to which rain penetrates is limited and salts become concentrated in the root zone. However, the addition of salts by irrigation water is more important. A continual supply of water in excess of that required by the growing crop and without adequate drainage results in a raising of the water-table to levels from which salts can be drawn by capillary water movement and evapotranspiration by crops with long-enough roots to tap this zone. Problems arise when the water-table level comes within about 2 m of the ground surface since at least 3 m depth is required to allow percolation and leaching to counteract the tendency towards salt accumulation. The *leaching factor* is the percentage of the total irrigation water applied that is required to wash excess salt out of the soil over a given period of time. No irrigation scheme can sustain agricultural productivity unless the water-table is kept at a low level and the soil remains permeable.

Irrigation not only tends to increase the total salt content but can, as has been indicated, lead to an imbalance in the proportion of component salts. With increasing alkalinity, consequent particularly on the increase in the sodium (>15 per cent total cation exchange capacity) and the bicarbonate-ion content, the physical condition of the soil starts to deteriorate. The process of *alkalinization* (*solonization*), in the narrow sense, is that whereby the cation exchange complex in the soil acquires an appreciable saturation with sodium ions. As long as free less-soluble salts are present, the pH remains below 8.5. Calcium exerts a flocculating influence on the clay fraction, which helps to create and maintain a loose crumb structure and maximum soil permeability. An increase in sodium at the expense of calcium ions can be the result, on the one hand, of the application of sodium-rich irrigation water or, on the other hand, of the leaching of calcium as a result of natural or deliberate soil drainage.

Table 13.5 Status of saline ('white alkali' or solonchak) and alkaline ('black alkali' or solonetz) soils

	Saline	*Alkaline*
Salt content	>2% free salt within 125 cm surface	no free salt
Exchangeable sodium percentage	<15%	>15%
pH	7.0–8.5	8.5–10.0

As the pH rises above 8.5 the deflocculating influence of the sodium affects the dispersion of the clay particles, resulting in the deterioration and the final breakdown of the soil structure. Soil permeability decreases. Further, the dispersed clay can be leached from the upper soil horizon to produce an impermeable B-horizon, which impedes drainage. Any organic matter present also breaks down very quickly into a black amorphous scum. Such alkaline soils, whether occurring naturally or as a result of irrigation, have been variously referred to as 'black alkali', solonetz, non-saline or saline-alkali soils. They differ from saline soils in an absence of free crystalline salts, an exchangeable sodium percentage of over 15 and a pH of 8.0–10.0 (see Table 13.5).

Irrigation without adequate measures to prevent salinization and alkalinization has resulted in the degradation and abandonment of land formerly cultivated. It has been estimated that in West Pakistan (Punjab) 15 per cent of the land brought into cultivation at the end of the nineteenth century by large modern irrigation schemes had been rendered uncultivable by salinization by 1960. Further, it was calculated that in Egypt the rise in the water-table had reduced the cotton yield by half its potential by 1945 (Balls, 1953). Currently nearly a third of the cultivated area has salinity problems; two-thirds have drainage problems. Prevention of salinization, however, requires careful management to ensure that the water used for irrigation is of suitable quality; that sufficient water is applied at the correct time and in the correct amount to effect leaching of excess salts; and that drainage systems are installed which maintain a sufficiently low water-table level to allow effective downward movement of water under the action of gravity. In addition, land-levelling is necessary to obviate the concentration of salts beneath 'high' spots and water ponding in 'low' spots. In some cases deep tillage is necessary to eliminate the variations in texture so characteristic of alluvial soils and where clay strata may create a 'perched' water-table. The plant is most vulnerable to salinity during germination and seedling development; hence management to avoid salt concentration is essential at this growth stage.

Most salt-management problems occur with *furrow irrigation* when salt accumulates on top or on the sides of the ridges in which crops are grown. *Sprinkler* and *drip irrigation*, both of which need relatively high-quality water to avoid foliage burn by sodium and chloride, help to reduce salt accumulation on the surface. Finally, timing of irrigation for salt control is very important. During the wetting cycle, salts are moved downwards with the depth of the wetting front (which is also the zone of highest salt concentration) depending on the depth of water penetration.

The ease with which salinized and alkalinized soils can be reclaimed depends on the extent to which degradation has proceeded. Provided the water-table can be lowered and the soil is permeable, salts can be removed by flushing. This can, however, result in the production of alkaline conditions in saline soils which do not contain gypsum ($CaSO_4 . 2H_2O$) and where there is a lack of a reserve of calcium ions to maintain the correct calcium to sodium ratio and ensure an exchangeable sodium percentage of less than 15. Measures to counteract alkalinity include the addition of gypsum and/or acidifying sulphates which can replace the exchangeable sodium cations. However, where alkalinization has been accompanied by the dispersion and leaching of the clay fraction, with the consequent formation of an impermeable B-horizon, rectification is extremely difficult, if not economically impossible.

IRRIGATION SYSTEMS

The basic purpose of any irrigation system is to convey water from a source to the cultivated field and to deliver it to the root zone of the crop. The practice of watering or irrigating land is very ancient. Many of the early civilizations of the Eastern Mediterranean area developed on the basis of irrigation agriculture. There are, as a result of a long period of evolution, a wide diversity of systems for transporting water and applying it to crop land (see Fig. 13.2). The less-sophisticated methods are dependent on intensive human and/or animal labour and are particularly characteristic of the less-developed tropical areas. Most of these involve the lifting of water from rivers, wells and, in some instances, small reservoirs, by hand or simply operated implements, and applying it directly on to adjacent fields or into open channels which feed nearby fields.

Modern, and usually large, irrigation schemes are dependent on the transpor-

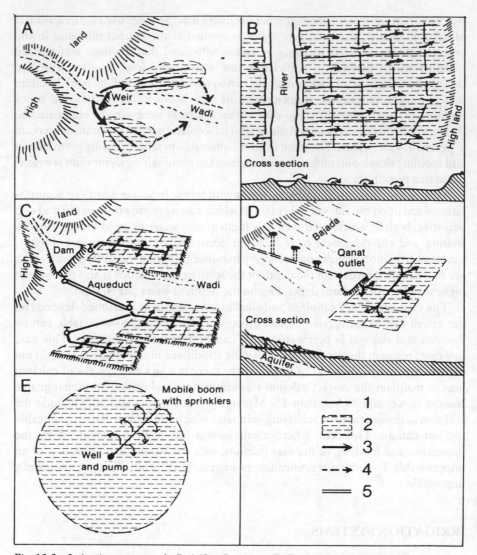

Fig. 13.2 Irrigation systems: A, Seil (flood) system; B, Basin system; C, Reticulation system (dam and canals); D, Reticulation system (*qanat* or *foggara*); E, Centre-pivot sprinkler system. Key: 1, dikes or walls; 2, irrigated areas; 3, water movement; 4, drainage of excess water; 5, canals (from Heathcote, 1983)

tation of water often considerable distances from high rainfall areas to large storage dams; on pumping ground water from great depths; and on applying water by one or other means depending on the crop, the water needed, infiltration rates and the water-holding capacity of the soil together with the land form.

Surface or *gravity irrigation* is that system whereby water distribution is controlled basically by the nature of the land surface. It requires land-levelling as a means of spreading water as rapidly as possible. Gradient is hence an important limiting factor. Spreading can be effected by means of 'wild' flooding, border checks and furrows. Gravity irrigation is one of the most widely used systems in both modern and traditional agriculture. It has the advantage that no pressure is required for distribution, and application of water is not affected by wind. However, it may result in the removal of more fertile top-soil when levelling is needed; and large but intermittent discharges of water do not make for most efficient use. Ponding and the formation of soil-surface crusting, together with high evaporative losses, are hazards on heavier soils. On lighter soils there may be excessive water percolation to below the root zone. More significantly, gravity irrigation tends to concentrate salt at the 'wetting front' – i.e. at the junction between wet and dry soil.

Sprinkler irrigation systems are characterized by the need for a source of water under pressure and by the use of pipelines for transport and nozzles for application. Fixed or movable pipes and rotating sprinklers are basic components. As the name implies, the water is applied as a spray which is comparable to rainfall. The method does not require land-levelling and can be used on rolling land. Drainage problems are less than under gravity methods and soil erosion is minimized. Systems vary according to the type of conveyance and of discharge device. Sprinkler irrigation has the advantage that it is easy to control the amount and rate of water application. It is particularly effective where (because of coarse and/or shallow soils, pans or a high water-table) small but frequent applications are required for the germination and establishment of young plants. Apart from the high costs of installation, evaporative water loss by wind from sprinklers can be high, while some crops can be damaged when leaves are wetted. Also, fungal and bacterial diseases are more prevalent. In Israel today 80 per cent of irrigation is by sprinkler systems.

Trickle irrigation has been developed more recently. It involves the discharge of small amounts of water from small orifices located on or immediately below the soil surface near the crop roots. It aims to make the most effective use of water while minimizing water losses, salinity hazards and surface soil disturbance.

Sub-irrigation is a more specialized system, which is still limited in use. It involves the management of the water-table level so as to bring the capillary water fringe within the root zone. It has, however, the disadvantage that it makes the soil susceptible to salinization because of the shallow ground-water level and the impossibility of leaching out excess salt.

Increasing efficiency of irrigated agriculture has been accompanied by the use of irrigation systems for the application of nutrients and pesticides and to modify microclimatic conditions. The latter now includes using water to cool the canopy, particularly of citrus crops, when temperatures reach dangerously high levels; to cool glasshouses and animal houses; and to provide winter protection to crops in areas prone to low temperatures. On the one hand, spraying can keep temperatures above freezing point; on the other, shallow solar-heated 'ponds' are used to protect young rice crops. Unfortunately, the increase in perennial irrigated agriculture has had a drastic impact on pest and disease incidence (Perfect, 1986). Large areas of standing water have encouraged the spread of a number of water-borne diseases such as malaria and schistosomiasis. In the case of pests the impact has been most severe in relation to those normally adapted to a markedly seasonal cropping regime. Under normal circumstances such pests have to survive an unfavourable period when the host plant is not available and then be capable of rapid reproduction when it is. When, as a result of irrigation, the seasonal constraints are removed, a pest population explosion can occur, with serious reduction of crop yield. This is illustrated by the case of the leaf-hopper (vector of the maize streak virus) which normally lives on seasonal grasses. It has become a serious pest on irrigated maize in Zimbabwe because the cultivation of winter cereals by irrigation has created an alternative habitat which maintains dense, relatively stable, populations of the leaf-hopper. The virus is now a serious cause of disease in irrigated maize. By the same token, irrigation systems have been manipulated successfully to control pests – such as the cotton-boll weevil in the USA – by microclimatic modification.

EFFICIENCY OF WATER USE

The efficiency with which water is used in irrigation is normally expressed in terms of the fraction of water withdrawn from rivers, reservoir or aqueduct that is *consumptively used*, i.e. used in evapotranspiration on the farm (see Table 13.6). A distinction is frequently made between the efficiency of:

1. Conveying water to the land.
2. Applying water to the land.

and together these can be as low as 30 per cent. The efficiency of transport is largely

Table 13.6 Amount of water diverted, delivered and used in twenty-two irrigation projects examined by the US Bureau of Reclamation

Water (surface cm³)					Efficiency	
Amount diverted (A)	Amount delivered to farm (B)	Evaporation (C)	Precipitation (D)	Amount consumed amount delivered (C – D)	Transport (B/A)	Application (C–D/B)
157	98	77	20	57	0.62	0.58

(from Marshall, 1972)

dependent on how far and by what means water is conveyed, since evaporative loss from open water surfaces in arid climates can be extremely high. The efficiency of application depends on the design of that system of irrigation best suited to the soil, water and crops. Among the factors that give high efficiency of water use and minimize evaporative loss are the depth and permeability of the soil in which water-table control can provide ample depth of rooting for a wide range of crops and a high *infiltration capacity* (i.e. the rate at which water can be accepted by the soil). The latter is highly variable in space and time since it is dependent on surface relief and the amount of water actually present in the soil, as well as on the type of crop and the amount of cover it provides. Infiltration capacity decreases as the soil becomes

Table 13.7 Effective root zone depth of some common crops grown in deep well-drained soils

Shallow (60 cm)	Moderate (90 cm)	Deep (120 cm)	Very Deep
rice	wheat	maize	sugar cane
potato	tobacco	cotton	citrus
cabbage	groundnut	sorghum	coffee
cauliflower	carrot	pearl	grapevine
lettuce	pea	millet	apple
onion	bean	soy	safflower
		sugar beet	lucerne

(from Michael, 1978)

wetter and may eventually attain a constant value dependent on the texture and structure (see Fig. 13.3). Soil water-storage capacity is also important since this will determine water loss in drainage, itself a function of soil texture and depth. The rooting depth (see Table 13.7) (i.e. the depth of soil in which 90 per cent of roots grow) must be sufficient to accommodate the rooting habit and depth of the particular crop.

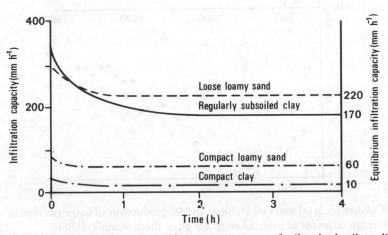

Fig. 13.3 Infiltration capacity and equilibrium on two types of soil and subsoil conditions (from MAFF, 1982b)

Considerable research has been conducted over the past 50 years to determine the water efficiency of crops under different growing conditions. Each crop operates within a characteristic range of evapotranspiration rates. The optimum rate is that which will ensure maximum yield under specific conditions of growth. The *optimum irrigation efficiency index* has been defined as the maximum value of the ratio of yield to seasonal depth of irrigation water applied under a given set of crop and environmental conditions. The traditional approach to increased efficiency by increasing yield without changing evapotranspiration need is now being superseded by efforts to increase yields by a decrease of water application to below the maximum evapotranspiration need. This is particularly well exemplified in Israel, where there has been a relatively recent and extremely rapid development of irrigated agriculture since the Second World War. In the early 1950s less than 20 per cent of the cultivated area was irrigated and this was mainly by ditch and basin systems to supply citrus orchards. Irrigation now accounts for 50 per cent of the cultivated area, an expansion achieved by the construction of the National Water Carrier completed in 1964. This links the Jordan River headwaters into the coastal aquifers and feeds large storage reservoirs which ensure a large constant supply of water in an area where mean annual rainfall is low and extremely variable.

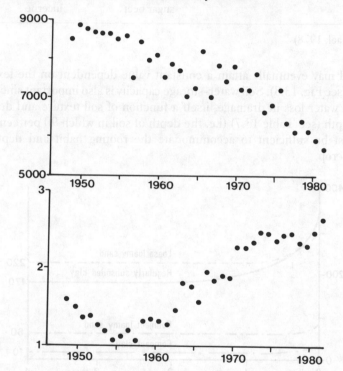

Fig. 13.4 Changes in (a) water expenditure and (b) productivity of irrigated crops in Israel. Average values for all irrigated crops are given (from Stanhill, 1986)

During the last 55 years or so the rate of water application per unit area irrigated has declined while productivity per unit area has increased (see Fig. 13.4). This increasing efficiency of water use has been effected by the application of the results of detailed research designed to determine the economically optimum irrigation regime for each major crop in each soil–climate regime (Stanhill, 1986); and to allow the uses of nutrient-rich brackish (Boyko, 1968) and sewage water for irrigation, among other uses. Today a fifth of Israel's irrigation water comes from these two sources. In addition, there has been a considerable development in the use of computers for the design and operation of irrigation systems and of remote sensing via landsat images to monitor crop water stress. These techniques have all contributed to the high and escalating cost of irrigation. As a result, the most recent research in the USA has been focused on ways of increasing the efficiency of the simpler less-costly gravity-flow systems and irrigation scheduling techniques.

While 'irrigation, probably more than any other factor, has contributed to the manipulation and to the increase of the productive potential of agriculture systems' (Perfect, 1986, 347), it has done and continues to do so at high economic, and often high environmental and social, cost. More water is transpired per unit of crop yield in arid than in temperate climates and the direct and indirect costs of supplying irrigation water can, particularly in arid and semi-arid regions, make it too expensive for the production of staple food crops.

14

Intensive agriculture

Modern intensive agriculture is characterized above all by the use of the most sophisticated technological methods of farming. It involves high levels of capital expenditure or inputs in order to achieve as high an output per unit of land area and/ or of livestock with the maximum efficiency possible. As well as fixed capital, and investment in land, buildings, livestock and machinery, intensive farming incurs large annual production costs. These include, on the one hand, those of supplying *direct energy* – in the form of human and/or animal labour, fossil fuels and electricity – necessary to undertake farm work; and, on the other, those of supplying *indirect energy*, represented by fertilizers, water, herbicides, pesticides, seeds and a wide

Table 14.1 Agricultural taxonomy based on energy (Gija-joules) density and protein yields

	Energy density (GJ ha^{-1})	Protein yield (kg ha^{-1})
Hunter-gatherer	0.0	
Andean village (Peru)	0.2	0.5
Hill sheep farming (Scotland)	0.6	1–1.5
Marginal farming	4.0	9
Open-range beef farming	5.0	130
Mixed farming in a developed country	12–15	500
Intensive crop production	15–20	2000
Fed-lot animal production	40	300

(from Slesser, 1975)

range of other chemical products necessary to obtain high levels of crop and/or livestock production. Slesser (1975) has used energy density or the input of total energy (direct and indirect) equivalents 'at the farm gate' per hectare of farmland as a means of expressing relative intensity of agriculture (see Table 14.1). The expenditure on the individual inputs varies with the type of farming (Fig. 14.1).

Temperate soft fruits (raspberries, strawberries etc.) are still very labour-intensive and these costs together with those for pesticides dominate the inputs in this case. In contrast, dairy farming is characterized by a higher use of electricity than are other types of livestock or arable farming, while arable cropping involves a proportionately higher expenditure on fuel oil, fertilizers and herbicides than do other types of farming.

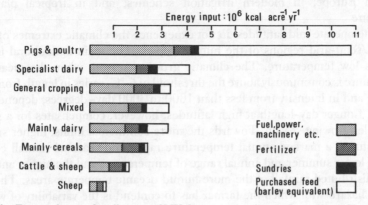

Fig. 14.1 Energy density for all types of UK farms, except horticulture (from Slesser, 1975)

Modern high-intensity agriculture has emerged only within the past 30–40 years as a result of closely related scientific-cum-technical developments which occurred during and immediately after the Second World War. These included:

1. Increasing mechanization of agriculture and the consequent replacement of energy formerly supplied by draught animals and manpower by the energy of the tractor, together with an increasingly diverse range of ever-more sophisticated implements designed to cope with every stage in the process of crop cultivation and livestock production. This process was stimulated by the shortage of agricultural labour as a result of the demands of the armed forces and wartime defence industries in both Western Europe and North America. In the mid-1940s there was still a high reliance on animal power in both the UK and the USA. Today there are few commercial field crops that are not capable of completely mechanized cultivation.
2. Increasing use of chemicals for fertilizers, insecticides, pesticides and herbicides, veterinary medicines and other non-essential additives.
3. The rapid developments in crop and animal breeding programmes, with the production of high-yielding crop varieties whose full potential could only be realized by a high input of nutrients and with a form and growth habit amenable to mechanized cultivation, and of livestock requiring an increasing proportion of high-energy feedstuffs to achieve high productivity.

Intensification of agriculture has, in addition, been accompanied by increasing specialization in the type of farming undertaken; spatial separation of crop and livestock production; a concentration of crop production on land of high capability for agriculture which will give the highest returns on capital invested; and an increase in the size of the production units in response to an increase in the cost of inputs at a rate greater than that of the price obtained for outputs. While intensive agriculture is not exclusive to any particular biophysical environment, it is most highly and extensively developed in the humid temperate regions of the world – particularly in the developed urban-industrial countries of North America and Western Europe; in modern irrigation schemes; and in tropical plantation agriculture.

The temperate mid-latitudes do not experience the climatic extremes of either the arid/semi-arid regions of the humid tropics. The main agricultural limiting factor is low temperature. The climatic growing season, when the mean daily temperature is continuously above the threshold of 6 °C, varies in length from 5 to 9 months, and in intensity from less than 1000 to 3000 day–degrees, depending on latitude. Longer day-length at high latitudes, however, compensates for a shorter and cooler growing season. Towards the more continental limits, higher summer temperatures, a marked annual temperature range and seasonal rainfall contrast with the lower summer and annual range of temperatures and the higher and more evenly distributed rainfall of the more humid oceanic temperate areas. The main climatic hazard with which the farmer has to contend is the variability of weather throughout the year, with risks of particularly late or early frosts; soil water deficits in summer in the more continental areas; and soil water surpluses in spring and autumn in the more oceanic areas. Physical conditions are, however, suitable for the cultivation of a wide range of crops and crop combinations.

Of the cereals, wheat, barley and maize (corn) are the most extensively cultivated. Wheat gives higher yields than in the drier prairie 'wheat lands'; and both hard and soft wheats can be grown in the warmer drier areas and in the cooler more humid areas respectively, though high rainfall can make the production of hard wheats marginal. In North-West Europe barley is now as widely cultivated as wheat and, with a short growing period, it has a more northerly range than any of the other temperate cereals. It has all but displaced oats and rye, formerly the traditional cereals grown under cooler and wetter climates and on poorer soils in temperate areas. The cultivation of autumn- and spring-sown varieties of both wheat and barley allows a degree of flexibility of cropping and recoverability of crop in the face of unseasonable weather conditions, though the higher yielding winter cereals are more prevalent. Quality, however, determines the ultimate use and economic value of these cereals. Bread wheat requires hard grains with a high gluten content; barley for malting an acceptably low water and, more importantly, low nitrogen content. In both instances, quality is highly weather-dependent. In Mid-Western North America and Central Europe, where summer temperatures and humidity are higher, maize (corn), combining high yields with a multiplicity of uses, is the dominant cereal.

Temperate root and tuber crops include the 'white' potato, and a number of *Brassicas* such as swedes (or rutabaga), turnips, mangolds and sugar beet. The *Brassicas*, particularly the swede (*B. napas*) and turnip (*B. campetris*) are an important source of winter fodder. The swede has a higher dry-matter content (10–12 per cent) compared with the latter (8–9 per cent), and is both frost-hardy and disease-resistant (Holmes, 1973). Its cultivars provide a spread of maturity dates from autumn to early spring, following spring to early summer planting. The white-fleshed turnip can produce a useful crop 10–16 weeks after sowing. It is, however, not so frost-resistant nor does it store so well as the yellow-fleshed type, which matures more slowly and stores better. However, the area under root crops has declined during the past 20–30 years because of increasing demand for fodder grain and the high labour requirements for the cultivation and harvesting of fodder roots. These disadvantages have, however, been mitigated to a certain extent by the drilling, instead of 'broadcasting', of the small seeds. This reduces the necessity to weed out surplus seedlings by hand – 'singling' – that was previously so labour-intensive.

The potato and sugar beet have retained their value as food crops. The white potato is particularly well adapted to the cool humid temperate climate and at the time of its introduction to Europe it was capable of producing more food per unit area on poorer soils and in wetter climates than were cereals. Potato yields are highest, however, on light rather than heavy soils; not only do the latter impede tuber growth but associated stoniness and a cloddy structure make mechanical lifting (harvesting) difficult. Maximum yields are attained where low night temperatures are combined with warm sunny days and with soil water near field capacity during the phase of tuber enlargement. In frost-free coastal areas early potatoes (harvested in May/June) can reach local markets about 2–3 weeks before those cultivated by irrigation in warmer Mediterranean regions. Because it is vegetatively propagated the potato crop is prone to several particularly severe leaf-virus diseases as well as to a number of bacterial and fungal infections of the tuber. As a result, potatoes grown for seed are normally cultivated on the cooler margins of their range, beyond that for optimum yield, and just on or beyond the optimum range of the disease-causing organisms.

The seed-legumes are, as in the Tropics, second only to the cereals in yield and area planted in temperate intensive farming systems. Among these the soya bean, the common field bean and the pea rank highest. They are particularly valuable multipurpose crops which produce food and fodder while, at the same time, contributing nitrogen and organic matter to the soil. In North America the soya bean (with similar climatic and soil requirements to those for maize) is a major source of vegetable oil. Its counterpart in Western Europe now is the long-maturing, high-yielding but non-leguminous oil-seed rape, which competes in acreage with wheat and barley in many intensive arable areas.

INTENSIVE ARABLE FARMING

The development of temperate arable agriculture has been accompanied by the replacement, over considerable areas, of the former traditional methods of grass (sod)/arable-crop rotation or ley farming by *continuous cultivation*. The former involved the rotation of cereal crops of varying tillage and nutrient requirements with a 'break' crop such as potatoes, fodder roots, soya beans or field beans, and the alternation of these arable field crops with a sown grass/legume sward (or 'ley') of approximately 3–6 years duration. This became the basis of the traditional mixed crop and livestock farming characteristic of humid temperate agriculture until the mid-1950s. The main advantage, indeed the essential role, of the grass/arable rotation, was that it combined a method of keeping weeds, pests and diseases (associated with a particular arable crop) in check with that of maintaining soil fertility. The ley grass, which was eventually ploughed under, received both live-stock droppings and top-dressings (mainly of lime and phosphate). It promoted a build-up within the soil of organic matter rich in nitrogen, and it contributed both directly through the action of the grass roots and indirectly through the activities of the rich soil fauna to the build-up of as stable a crumb structure as possible given the soil texture (Page, 1972). In addition, the root crop, which was also an important component in this traditional rotation system, provided opportunities for weed control in inter-row cultivation as well as for control of specific soil-borne diseases and pests by crop diversification (Holmes, 1973). Finally, the rotation system had the advantage of spreading labour and machinery requirements throughout the year.

Since the 1950s, crop rotation systems have gradually been modified or relin-quished in the face of the increasing use of inorganic fertilizers and the development of high-yielding crops which need greater nutrient inputs than their predecessors; the replacement of labour and draught animals with better, and increasingly bigger, machines for cultivation and harvesting – of which the combine-harvester is one of the most important – and the increasing use of herbicides, pesticides and growth regulators. To these factors must, of course, be added the increasing demand for cereals, particularly for animal feedstuff, together with guaranteed-price support mechanisms. The maintenance of soil fertility is no longer dependent on a particular crop rotation system and the use of FYM, while increasing speed of operations, consequent on mechanization, allows greater flexibility in the choice of crops and cropping systems. Among these changes have been the decline in production of the fodder root crop (the traditional winter-feed stand-by) with a proportionate increase in potatoes, sugar beet and root vegetables (e.g. carrots); the replacement of clover or clover/grass by field beans and vining peas; and the cultivation of one particular cereal crop for several (2–4) successive years, instead of an alternation of one or two cereal years with a 'break' crop (Lockhart and Wiseman, 1983).

Continuous arable cultivation has, to a greater or lesser degree, replaced the grass/crop rotation, particularly in those areas best suited climatically to arable cereal farming. Simplification of crop combinations has been accompanied by increasing crop specialization both regionally and within the farm unit. Cereals have played a central role in the intensification of arable agriculture, as the means of increasing crop yields became available in the years immediately after the Second World War (Fiddian, 1973). The development of increasingly high-yielding varieties of maize, wheat and barley was accompanied by a growing need for and use of inorganic fertilizers. The early rapid root growth and deep rooting of the new cultivars were particularly demanding of nitrogen. New slow-release forms of nitrogen fertilizers or nitrification inhibitors were developed with the aim of reducing losses between application and uptake by the crop. The inhibitor *nitropytin* doubled yield response of winter wheat to autumn applications of nitrogen in Indiana. In addition, the production of cereal varieties with larger heads but shorter and stiffer stems allowed high nitrogen input without the risk of crop loss by 'lodging' (i.e. falling over because of the weight of the plant).

Temperate cereals can now be chemically manipulated at various growth stages by plant growth regulators (PGRs) (Batch, 1979). Until relatively recently these have been used mainly in the production of horticultural and ornamental crops, e.g. for chemical pruning; joint-loosening and ripening to aid harvesting; and controlling fruit-set in tree crops. Research has now started to focus on the use of PGRs to increase yields in arable crops in response to the increasing costs and environmental problems of plant-breeding programmes. In 1979 several commercial products became available which could be used to promote development of short strong cereal stems. So far the most successful and widely used PGR is *chlormequat*. However, although developed in the 1960s it has only been accepted by some countries in Europe (UK and Germany). As Batch notes, 'the benefits of chlormequat and other PGRs are only manifest under high input systems as an interaction with other inputs' (1979, 374).

Although the straw to grain ratio has decreased in the new HYVs of wheat and barley, there has been an increase in straw production consequent on the extension of the acreage under cereals. Nevertheless, the amount harvested in England and Wales has declined markedly; and 40–50 per cent of straw is now burned in the field after the crop has been harvested. Decrease in the use of straw is a result of the decline of livestock numbers in the most intensive cereal-growing areas and its low value in relation to transport costs. *Stubble-burning* provides a quick and cheap method of cleaning the ground after combining in order to facilitate the establishment of the next crop before the winter. It also helps to depress weeds, pests and disease organisms while giving a quick release of phosphorus and potassium. It has now been established that burning does not have any serious long-term effects on soil organic-matter content since roots, chaff and stubble remains contribute a much higher proportion than does cereal straw (Staniforth, 1974). Fire and smoke hazards, however, can be high and in the UK the National Farmers Union has drawn up a code of practice to minimize these risks.

Increasing specialization in grain crops has resulted in the resurgence of the weed problem. All cereals are, to varying degrees, soil-exhaustive and, if grown continuously on the same land for too long, they will be accompanied by an increase in weeds and a build-up of associated fungal diseases in the soil. The 'corn-borer', one of the most virulent pests of maize, has attained pest status in the USA only within the last 40 years. Fungi responsible for 'take-all' and 'eye-spot', and pests such as eelworm in wheat and barley, can build up in the soil over a period of time and seriously reduce yields. The discovery of the growth-regulating herbicides MCPA and 2.4D in 1942 provided an effective chemical method of widespread control of most broad-leaved and, particularly, annual weeds. Varying resistance to these first herbicides, however, also created new weed problems. Some more-resistant plants such as cleavers (*Galium aparne*) and mayweed (*Matricaria chamonilla*) and perennial grass weeds increased. Of the last, the common wild oats (*Avena fatua*) underwent an epidemic increase in the later 1960s; it was joined by winter wild oat (*A. ludoviciana*) and black-grass (*Alopecurus mysuroides*), which germinate in the winter and consequently infest winter cereals. The grass weeds not only cause more crop damage and loss but they are more costly and more difficult to control than other types of weeds. This is because they have a short life-cycle, mature before the host crop, and have a seed, which, by 'burying itself' in the soil, can remain viable for several years – commonly up to 3 years but even up to 9 years in exceptional cases.

Similarly, intensive cereal cultivation has generated serious and sometimes new pest and disease problems. Animal pests, particularly birds such as the rook and pigeon, have also increased in number. Of more serious significance has been the proliferation of pathogenic organisms, which has been exacerbated by high planting densities and by monocultures and continuous cultivation of cereals. As Haines (1982) points out, the same pattern of intervention year after year discriminates in favour of some pests and diseases as it does for weeds. The control of pests and diseases is difficult and complicated. With increasing intensity of farming has gone an increasing use of chemical pesticides; and it is now common practice to spray for this purpose several times during the growing season. Breeding of new disease-resistant varieties is expensive and takes time and its success tends to be offset by the rapid evolution of more virulent and chemically resistant pests.

The inevitable need for a break in cultivation has seen the adaptation of rotation systems to the requirements of intensive agriculture. In some cases this involves the introduction of a high-value cash crop where soil conditions are suitable, or of a short ley. More common are crop combinations which give a cereal break but where cultivation is compatible with that of the cereal in that the crops can be grown, harvested and dried with the same machinery. These include field beans and peas, seed crops of grass, clover and potatoes and, more recently, linseed and oil-seed rape in the UK; and soya beans and mixtures of soya beans and maize in the USA. In addition, the introduction of these non-cereal crops does not require livestock to be kept; they improve the soil; and labour requirements for cultivation are complementary (Haines, 1982).

Continuing intensification of arable farming has been accompanied by an increase in the number and size of machines. While timeliness, speed of operation and the condition of the harvested product are thereby improved, mechanization has created its own problems. Larger field sizes (with the elimination of former boundaries) and farm units are necessary to accommodate the size and cost of more and bigger machines. Continuous cultivation has resulted in an increasing intensity of machine-tracking to which an arable field can be subjected in the course of the year. This has resulted in overcultivation, which has, in some areas, caused a serious deterioration in the condition of the soil. Plough pans, resulting from compaction or smearing at a constantly maintained ploughing depth, develop and disrupt soil drainage. The tracking of heavy machines also causes surface compaction with a consequent decline in the bulk density, permeability and infiltration capacity of the soil. Without the use of FYM, the organic-matter content can drop to minimum levels (*c.* 2 per cent) and the structural stability of the soil be destroyed. Under such conditions the risk of soil erosion either by surface run-off on the heavier or by blowing on the lighter soils increases.

In order to mitigate the deleterious effects of continuous cultivation some farmers have turned to *reduced* or *minimum tillage* (i.e. *zero* or *no-till*) techniques, which involve seeding into unploughed soils. This technique has been introduced and adopted to a greater or lesser extent in temperate Europe and North America. Crops are drilled directly into the soil after the surface vegetation has been killed off with paraquat. In Southern England zero tillage was initially used for seeding kale into a former ley; it was later extended to other Brassicas, oil-seed rape and fodder maize and to seed winter wheat into the stubble of the previous crop. In the USA it is used to seed maize into a ley. The successful adoption of no-till seeding has been dependent on the development of new heavy, deeply penetrating, drills, as well as on the availability of light well-drained soils, free of perennial weeds. However, it does not entirely eliminate the need to plough; this is still necessary for the cultivation of root crops and may be necessary for cereals in some seasons. In other cases minimum tillage may be accompanied by 'tramlining', the provision of permanently unsown tracks through a crop, along which machine movement and hence pressures are concentrated. The use of larger wider tyres also helps to spread the weight load and mitigate soil compaction.

INTENSIVE GRASSLAND FARMING

In comparison to the major temperate field crops presently cultivated the cost per acre of establishing and maintaining grass and forage legumes is high and returns are relatively low. In addition, the increasing use of inorganic fertilizers, herbicides and pesticides has seen a decline in the value of the grass ley as a 'break' crop and as a source of livestock feed in many intensive farming systems. On the other hand, in temperate areas where environmental conditions are less suitable for seed crops (i.e. cereals and legumes), the grass/legume crop has become increasingly important as an alternative feedstuff to grain and as the basis of livestock and, particularly, intensive dairy farming.

The agricultural potential of grass is high. Once established it is a relatively cheaply produced feedstuff that can be managed at different levels of input and intensity to suit varying economic and ecological conditions. Where mean daily temperatures exceed 5 °C and soil moisture is not limiting, grass can continue to grow; and, with an LAI of 2–6, it can provide a complete ground cover throughout the year in temperate regions. Further, this cover can be maintained while the sward is being subjected to repeated defoliation by grazing and/or cutting. All grass species possess two meristematic zones, one at the base of the leaf blade, the other at the base of the leaf sheath; and the leaf matures from the tip downwards (Fig. 14.2). Defoliation stimulates leaf growth and the activity of axial buds which produce new vegetative shoots or 'tillers'; tillering ability, however, varies with species as well as with management.

Most cool-temperate grasses have low temperature thresholds for growth and optimum temperatures of 18–28 °C for leaf appearance. Perennial varieties, however, tend to be thermoperiodic and/or photoperiodic, needing low winter temperatures (0–10 °C) and/or long day-length for successful reproduction. The exceptions are annual grasses such as varieties of hybrid ryegrass (Westerwolds and

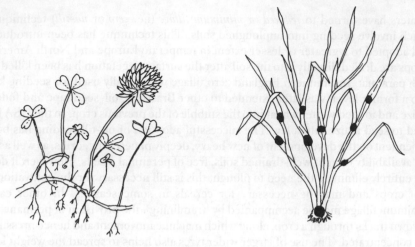

Fig. 14.2 Typical white clover and grass plant, showing positions of meristems (from Morris, 1977)

Italian), *Poa annua* and timothy (*Phleum pratense*). Grass, however, requires a constant water supply for maximum growth. Water deficits of 40–50 mm in the top 300 mm of the soil will retard growth; and although the deeply penetrating root system can tap water from depths down to 600 mm, a deficiency of available nitrogen below *c*. 200 mm becomes a serious limiting factor (Holmes, 1973).

The selection and breeding for agriculture of HYVs of grass is, compared to that for the cereals and other field crops, a relatively recent development. Many of the major advances in grass breeding have been achieved only within the last 50–60 years. The two main results of these breeding programmes have been:

1. The production of grasses in which the leaf blade comprises a large proportion of the total plant (i.e. the blade to stem ratio is high) and in which the C : N ratio is low; hence the nutritive value of grass as a feedstuff is high.
2. The concentration of breeding within a few major genera.

These include: perennial ryegrass (*Lolium perenne*), comprising a very large number of varieties – now the most important agricultural grass in Western Europe. Easy to establish, it is a persistent and an aggressive competitor in the face of weeds and other grasses. It has a high yield potential in response to nitrogen fertilizer and a digestible value higher than any other perennial grass. In contrast, Italian ryegrass (*L. multiflorum*) is short-lived, but is one of the earliest grasses to commence growth in the spring. Timothy (*Phleum pratense*), a traditional hay crop, is very palatable and winter-hardy, while cocksfoot (*Dactylis glomerata*) has a low digestible value, and tends to become coarse and unpalatable if undergrazed; it performs best under rotational grazing. Two other grasses are often used in mixtures: meadow fescue (*Festuca pratense*) and tall fescue (*F. arundinacea*). The former has a low competitive value; the latter is coarse-leaved but has a higher digestibility than cocksfoot and a rapid drying ability.

The grass crop is normally composed of a mixture of varieties of one or more of these species combined with a leguminous forage crop of which the most commonly sown in temperate regions today are the clovers (*Trifolium spp.*). The commonest, white clover (*T. repens*), is a very persistent species, tolerant of a wide range of environmental conditions but, sensitive to soil water deficits, it is better suited to wetter areas. Its digestible value is very high (72–75 per cent) and its nitrogen fixation is estimated at 135–200 kg ha^{-1}. Red clover (*T. pratense*), in contrast, is relatively short-lived (18 months to 2 years) but gives high yields (three cuts per year are possible) and a high protein content. Grown alone or mixed with grass, it is normally conserved, being well suited to silage. Other clovers include alsike clover (*T. hybridum*), native to Sweden, and well adapted to cool or cold climatic conditions, with high rainfall and acid soils of low fertility; and crimson clover (*T. incarnatum*), common in the USA.

Lucerne or alfalfa (*Medicago* spp.) comprises two species: *M. sativa* with an upright habit, which can grow under a wide range of temperature regimes in Mediterranean to cool-temperate regions on lime-rich soils; and *M. falcata*, with a prostrate habit, which is cold-tolerant, frost-resistant and can grow on acid soils. Lucerne has a deeply penetrating tap-root, which imparts drought resistance. While it can be grazed, it is mostly used for hay. It is very productive on light soils and in dry climates, where it gives higher and more nutritious yields than any other forage plant. Its digestible protein content is twice that of clover and four times that of corn (maize) silage. It is now one of the most important forage crops in the USA, where it has displaced clover in grass mixtures.

Like the grasses, the legumes can also maintain growth in the face of repeated defoliation. Most of the growth points are in the axillary buds located very near ground level, and the growth of the main shoot and of short shoots is stimulated by the removal of apical buds by either grazing or cutting. White clover can reproduce vegetatively by creeping stems (stolons), while the other legumes possess penetrating tap-roots. All the forage legumes possess the ability to fix nitrogen as a result of a symbiotic relationship with a bacterium (*Rhizobium* spp.) which lives in the root nodules. The association is, however, dependent on the presence in the soil of the strain of *Rhizobium* specific to the particular legume, which in turn, is dependent on the soil environment – particularly on there being adequate drainage and nutrients for the requirements of the bacterium. Soil acidity is one of the main limiting factors which results in a deficiency of free-living or symbiotic nitrogen-fixing bacteria. The inclusion of legumes with grass produces a forage with a higher protein content and hence higher digestibility than that of grass alone. However, the competitive ability of the legumes is lower than that of the grasses and their growth is depressed by the vigorous grass growth consequent upon high inputs of nitrogen fertilizers.

The composition of the grass/legume crop varies according to the mixture of species selected for sowing (see Table 14.2). The grass component, in particular, will be selected for feed value; earliness or lateness of production during the growing season (see Fig. 14.3); stiffness, particularly of leaf and stem for hay/silage making; and growth habit – prostrate or upright in relation to the grazing habits of the domestic livestock. The mixture chosen will be that which will best fulfil a particular use requirement under the available site conditions, at an acceptable cost, for as long as possible (Haines, 1982). Further, in modern intensive grass-based farming

Table 14.2 Typical seed mixtures

	Seed rate (kg ha^{-1})						
	Mixture no. *						Pure sowing of species
	1	2	3	4	5	6	
Italian ryegrass	14	34	–	–	–	–	25–35
Perennial ryegrass	–	–	20	–	10	–	16–22
Timothy	–	–	–	7	4	(4)	10–24
						or	
Cocksfoot	–	–	–	–	5	(4)	15–20
						or	
Meadow fescue	–	–	–	14	–	(4)	16–20
Tall fescue	–	–	–	–	–	–	14–20
Red clover	9	–	–	–	4	–	10–14
White clover	–	–	2	2	2	–	–
Lucerne	–	–	–	–	–	16	16–18

* 1 One-year ley; 2, 18- to 30-month ley intensively managed; 3, medium-duration ley 3 years or over; 4, suitable for dairy farms; 5, older Cockle Park mixture.

(from Williams, 1980)

systems the aim is to achieve as high a production of palatable nutritious forage (for grazing) and/or conserved feedstuff as will support the maximum output of livestock products.

Increased grass productivity has been won by a combination of high inputs, particularly of nitrogen fertilizer, and grass varieties with high yield potentials; but at the cost of depressing nitrogen fixation in the soil. This has been justified by the fact that nitrogen fixation in early spring is considered insufficient to meet the needs of the growing grass. In addition, where there may be a risk of seasonal water deficits of over 50 mm during the growing season, supplemental irrigation is commonly used, particularly if it can also be employed for another highly productive crop such as field beans, oil-seed rape, potatoes etc. To realize the productive potential of this type of grassland requires a high degree of management of the grazing and cutting regimes respectively. Low-level grazing management usually results in selective *undergrazing*, as a result of which the most nutritious grasses are eaten and the less nutritious neglected; the grass sward does not attain its maximum productivity and less than 50 per cent of the available annual production is consumed. Intensive grassland management aims to ensure as high a grazing level as is compatible with the maintenance of the productivity of the sward and without a degeneration in its condition either by overgrazing and/or surface poaching.

The main grazing systems which have been introduced in many cool-temperate areas in order to make a more intensive use of grass production are:

1. *Intensive continuous stocking*, which keeps the stocking rate in balance with grass

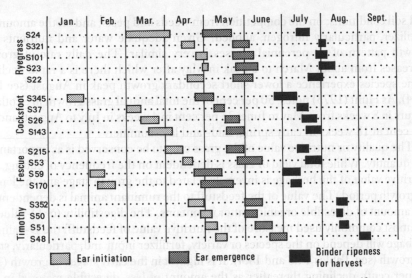

Fig. 14.3 Inflorescence development in selected grass species and varieties (from Spedding and Diekmahns, 1972)

production and utilization in the early part of the season and later, when sward production decreases, either reduces the number of stock or increases the area of grazing available in order to maintain the initial numbers. Continuous stocking encourages the development of a well-developed sward with a high proportion of clover.

2. *Rotational grazing*, which, as the term implies, involves the alternation of grazing at a selected rate for a given period of time with a rest period within fenced plots or paddocks. A variation of this procedure is *strip grazing*, by which stock are given access to a fresh allocation of grazing each day.
3. *Forage feeding* (so called 'zero-grazing' or 'mechanical grazing') has, for a variety of reasons been adopted on some farms in place of direct grazing. Grass is cut at its most productive and nutritious stage and taken to the animal daily for immediate consumption.

Although grass may, in many temperate areas, be able to grow throughout the year, it can only supply sufficient grazing to maintain livestock for 5–9 months. Some of the grass crop grown on the farm must, therefore, be conserved to provide winter feedstuff. The proportion of the maintenance and production needs of the livestock supplied by the farm's grass crop will depend on the quantity and quality available and the way in which it is conserved.

The yield of a grass which is being 'cropped' during its growing period is difficult to measure since growth can continue from the leaf meristems after grazing and cutting. However, the resulting production will nearly always exceed that of unexploited grassland grown under comparable conditions. It is generally considered that maximum yield in the UK is achieved when grass is cut at 4-weekly intervals during the growing season (Hunt, 1973). The maximum level, however, will depend on the particular mixture of grasses and legumes sown down; on the local climate and soil conditions; on the competition from weeds and pests; and on the amount of fertilizer, particularly nitrogen, applied. Given adequate water and nutrients the growth pattern of all perennial grasses is very similar. The daily rate of growth increases from about March to a peak in May after which there is a rapid decline; some species experience a lower short secondary growth peak in August (see Fig. 14.4). As Hunt (1973) notes, 50 per cent of the total annual grass growth in Scotland occurs in 8 weeks from May to June; 25 per cent in 4 weeks in July to August; and 25 per cent in 18 weeks in early spring, mid- and late-summer.

The quality or nutritive value of the grass crop to be conserved is as important as the quantity. While total yield of grass is dependent on the frequency of cutting, the nutritive value of the herbage is inversely related to the growth stage and the length of growth period. The value of the feedstuff to the ruminant animal is dependent on the amount of digestible organic matter it contains. The digestibility (or D-value) is the most commonly used measure of the nutritive value of feedstuff: the D-value of herbage will depend on the species or variety, fertilizer input and, particularly, stage of growth (see Tables 14.3 and 14.4). It is highest in the early stages of growth (74–77 per cent), declining thereafter as the amount of less-digestible material in the stem and cell walls increases, with the final nutritive value of the crop dependent particularly on the leaf to stem ratio at age of use. Decline in nutritive value starts with the stage of flower (ear) emergence, the earliness of lateness of which is a function of the particular species or variety of grass/legume of which the sward is composed.

Digestibility falls at *c.* 0.5 per cent per day during flowering within a range from 85 to 40 per cent; when lower than 70 per cent the increasing accumulation of undigested residue in the stomach and intestines reduces livestock appetite and feed intake. The feeding value of the conserved grass crop is, therefore, dependent on: the nutritive value of the original herbage; the growth stage at cutting; the proportion of legumes in the sward; and the amount of fertilizer applied during growth. The dilemma of grass conservation is 'how to achieve the best compromise between yield and quality' (Murdoch, 1980, 171). *Hay*, which is cut and field-dried (or cured), is usually harvested after the stage of maximum nutritive quality but when total yield of dry matter is highest. This is partly because cuts from an older upright sward are easier to dry than those from a new leafy one and partly because of the dependence on weather conditions for successful drying. *Field-dried hay* tends to be of relatively low and variable feeding value not only because of the growth stage at which it is cut but also because of a decline in its value due to 'weathering', particularly when drying is protracted or inefficient because of weather conditions. Good hay production requires several consecutive days of dry weather. The grass must be relatively dry before cutting, and drying must reduce moisture content of the cut to less than 25 per cent. While Britain may have an ideal grass-growing climate, it has one of the least dependable for hay making. Therefore, nutrient loss in the field and while in storage can be high. Field-dried hay can, at best, only provide the sort of feed that will maintain the animal's body weight and condition. A more 'concentrated' form of feed is necessary to supply production (meat or milk) needs. In contrast, *barn-dried* and, particularly, artificially *machine-dried* hay retains a much higher nutritive value but is relatively expensive to produce.

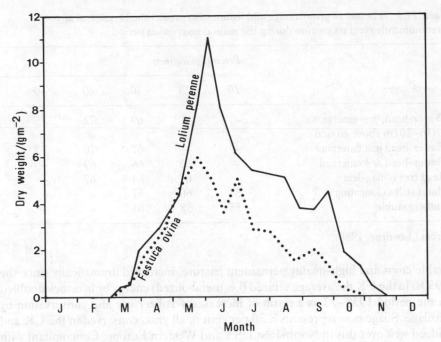

Fig. 14.4 Daily growth rates of sheep's fescue (*Festuca ovina*) and perennial ryegrass (*Lolium perenne*) on lowland sites (from Morris, 1977)

Table 14.3 Approximate D-value of leafy regrowths of swards of grass species at 4 and 6 weeks after cutting

Grass species	4 weeks after cutting	6 weeks after cutting
Perennial ryegrass	74	70
Timothy	72	68
Tall fescue	70	66
Cocksfoot	67	60
Italian ryegrass	66	62

(from Osbourne, 1980)

Silage, produced from grass and other green-plant materials, is a much more concentrated nutritive feedstuff than hay and one which can satisfy the winter production needs of livestock. Grass for ensilage is normally cut in the early nutritive stages of growth and, with adequate moisture and nitrogen input, may supply two or three cuts in the growing season, depending on environmental conditions. In addition, silage differs from hay in that it contains additives necessary to the process of fermentation and chemicals produced by enzyme action during ensilage. The productivity of the intensively managed rotation grassland has, like that of other

Table 14.4   The use of growth stage and leafiness to predict the D-value of any predominantly ryegrass pasture during the main conservation period

Growth stage	Percentage leafiness				
	10	*20*	*30*	*40*	*50*
Flower-head, pre-emergence (16–20 cm above ground)	–	–	69	72	75
Flower-head just emerging	–	–	67	70	73
Flower-head 3/4 emerged	–	63	66	69	–
Head free of flag-leaf	–	61	64	67	–
Floret stalks elongating	57	60	62	–	–
Anthers visible	55	58	61	–	–

(from Osbourne, 1980)

arable crops and high-quality permanent pasture, increased dramatically since the 1950s. In the UK the average utilized (i.e. metabolized) energy by livestock doubled in the period 1938–73 as a result of increased fertilizer use and conservation by ensilage. Silage now represents *c.* 50 per cent of all grass conserved in the UK and indeed well over this in North-East USA and Western Europe. Concomitant with this trend have been higher stocking rates.

The extent to which grassland farming has been intensified varies. As with arable cropping it has been most marked in areas where environmental conditions are relatively more favourable. It also varies between countries: the productivity of UK grasslands has not reached the same level as those in Denmark or the Netherlands, where nitrogen application has been higher. Indeed it has been estimated that the potential productivity of British grassland is considerably greater than the current average national yield. Also, utilization is limited by undergrazing: on average no more than 60 per cent of the growth is grazed, while maximum levels could be 75–80 per cent.

SWARD DETERIORATION

Improved grassland – be it temporary or permanent – tends once it has reached maximum yield to deteriorate in productivity at a rate dependent on whether it was established as a short- or as a long-duration ley. Some 'permanent' pasture on soils of high fertility can, however, be maintained at a high level for a long period if judiciously managed by grazing and fertilizing. Otherwise, sward deterioration is reflected in a decline in the proportion of clover, ryegrass and other sown grasses to non-sown 'weed grasses'. Where nitrogen inputs have been particularly high and cutting has exceeded grazing, a loss of available potassium from the soil may result. Also, overstocking in winter, when the soil is too wet, can result in compaction and puddling of the surface, breaking the sward and reducing infiltration capacity.

Increasing intensity of grassland farming has exacerbated these impacts as well as causing new types of deterioration resulting primarily from the increasing use of nitrogen fertilizer. These include 'winter kill', a condition which has become very prevalent in the past two decades in the West of Scotland. Bleached, and apparently dead grass appears in spring and, although the sward may recover by early summer, there is a deficiency of a spring bite for livestock (Hunt, 1973). 'Sward thinning' is another problem, particularly where $300 \, \text{kg} \, \text{N} \, \text{ha}^{-1}$ is applied annually to produce as large a hay/silage cut as possible. Growth after cutting tends to be slow and weeds such as couch grass (*Agropyron repens*) and chickweed (*Stellaria media*) can take a hold. They are difficult to eradicate and depress subsequent yields. Thinning also takes place as a result of intensive grazing, which leaves a dense build-up of grass short-bases open to weed infestation. In addition, the rapid leaf growth promoted by high nitrogen input is accompanied by a reduced and shallow root system, which gives a loose turf easily pulled out by grazing cattle. Finally, intensive grassland farming can have the same effect on the build-up of plant pathogens (fungi and viruses) and/or organisms parasitic on livestock. Haines (1982) notes that losses due to fungi and viruses can be as high as 20 per cent dry matter of the grass crop. Grazing regimes need to be managed to reduce infestation of parasites and worms. Those such as liverfluke (*Fasciola hepatica*), gut and tape worms and *Dictyocaulus viviparus*, which causes parasitic bronchitis, have two-stage life-cycles, one of which is outside the animal. Hence swards need a rest of at least a year from use by the same class of livestock to reduce the parasite densities.

INTENSIVE LIVESTOCK FARMING

Parallel with the intensification of cropping in temperate regions has gone that of livestock production, the aim of which is to obtain the maximum output possible of livestock products (meat, milk, eggs) from a given input of feedstuff. This has been, and continues to be, dependent on scientific and technical developments in livestock feeding; livestock management; and management and control of the animal environment.

FEEDING

The traditional methods of livestock farming, generally widespread until the 1950s, were based on the use of low-quality fodder, i.e. hay, grass and roots. Increase in animal productivity has been achieved primarily by an increasing use of concentrated high-quality high-energy feedstuffs, such as cereals, oil-seeds and silage. This has been accompanied by a marked improvement in efficiency of feed use as expressed in high conversion ratios (fodder eaten to food produced) and in the higher stocking rates that can now be maintained than formerly. In the UK, intensification of livestock farming commenced in the early 1960s with the rearing of purchased calves from dairy herds and the replacement of suckled feeding by milk substitutes. However, the extent to which high-intensity feeding has developed varies. Intensive livestock production can be part of an agrosystem in which only part of the production is intensive, as in the case of 'finishing' animals; or of one in which the whole production is intensive, as in the case of dairy, pig and poultry systems.

Sheep are still the least-intensively managed livestock. Intensification of beef and dairy production has been dependent on grass providing a high proportion of the summer diet and an increasing proportion of winter feed in the form of silage supplemented by concentrates. Short-term fattening on highly concentrated diets is characteristic of the corn-fed beef in the USA and of the barley-beef in Europe. The ultimate development of this trend is the American *feed-lot* of the South-Western states. Up to 50 000–100 000 head of cattle (bought in as yearlings) are kept at high density in open pens, and fattened, on a 6-month cycle, on a diet of grain and sugar-beet silage brought in from surrounding farms. In contrast, pigs and poultry are non-ruminant omnivores whose traditional role in the less-developed agro-ecosystem is still that of 'scavengers' and opportunistic feeders on farm wasteland or waste products. Intensive production of both is dependent almost entirely on a supply of concentrated feedstuff (e.g. maize, oil-seed, barley etc.).

Increased nutrition has allowed not only higher production of animal protein but higher productivity in terms of output per animal per unit time. Further, in highly controlled livestock systems feeding can be carefully regulated in order to ensure a desired production within a given time. In addition, hormones such as oestrogen are used to promote rapid weight gain in meat-producing animals in the U.S.A. However, increase in the nutritive value of feedstuff and intensive-feeding methods have in certain circumstances created dietary imbalances or deficiencies which must be made good if high efficiency of use is to be maintained. Nutritive problems can arise because of differences between the amounts of minerals required by high producers, such as dairy cows, and those present in the grass forage – including manganese and, in some areas, copper and iodine. In the case of pigs and poultry, which, under 'open-range' feeding, obtain their dietary requirements from a wide range of natural plant and animal sources, concentrated feedstuffs can be lacking in essential minerals and vitamins. In addition, livestock bred and reared under completely controlled indoor environments are subject to high stress levels, which can make the nature of their diet even more critical for high performance. Other feed additives can include antibiotics, when disease risk is high in young stock, and pigments designed to make the end-product as attractive as possible as in the case of the bright-orange egg yolk.

LIVESTOCK MANAGEMENT

Greater efficiency of feedstuff used and high conversion ratios have also been dependent on the great advances in animal husbandry that have taken place since the Second World War. Genetic research combined with intensive breeding programmes, aided by the now widespread use of artificial insemination, have produced livestock with potentially high productive capacities and an ability to metabolize high-concentrate diets. Breeding capacity has also been improved, as in the case of pigs and poultry, the latter producing a larger number of eggs per hen-life. Similarly the more recent trend towards intensification of sheep farming has been towards increasing lambing incidence (e.g. three in 2 years) and the percentage of twins or triplets per gestation. Manipulation of breeding habits and life-cycles has contributed to year-round milk and meat production. Finally, advances in veterinary medicines have aided disease control, which has assumed an increased significance as stocking rates and animal housing densities have increased.

ENVIRONMENTAL CONTROL

Intensification of livestock farming has been accompanied by the development of controlled environments for part or all of the production process. Nearly all commercial pig and poultry production is indoors. The so-called 'factory-farm' was pioneered by the broiler industry in the USA, which superseded the 'luxury roaster' production after the Second World War. The adoption of intensive poultry housing was stimulated by the need to achieve economies of scale which would counteract the decreasing profits per bird from the traditional open-range system. This move allowed the manipulation of breeding and egg production by controlled lighting and the maintenance of an environment in which not only light but temperature, humidity, atmospheric dust and noise levels could be controlled. Increase in house size (i.e. batteries which could accommodate 1000+ head of laying birds) facilitated the complete mechanization of the preparation and mixing of feedstuffs; of food and water distribution; and of manure disposal. The mechanization of milking in the dairy industry has resulted in the development of milking parlours of increasing sophistication while allowing reduction in labour required to that of one man for every 70 cows or more. The production of twin and triplet lambs has necessitated housed lambing.

CONCLUSION

Intensification and increasing productivity of agriculture has been achieved by increasing inputs, of which plant and/or animal nutrition is the most important. High crop yields are dependent on large fertilizer, particularly nitrogen, inputs, and high animal productivity is dependent on feed of high nutritive value. Both, however, are achieved at very considerable energy costs and decreasing energy efficiency in terms of the ratio between energy used and food energy produced. Further, increasing animal protein production is being achieved at the expense of using high-quality food for animal feedstuff. More protein is fed to high-producing livestock than can be recovered from them. In fact, it has been suggested that breeding and selection has produced a few specialized varieties of livestock that are dependent on high-cost high-energy feed and that would be incapable of surviving on low-cost low-energy feedstuff. This is reflected in selection against stomach capacity; the rumen accounts for *c.* 18 per cent of the body weight of the average British dairy cow, in contrast to 30 per cent of that of a Third World cow! Not only has intensification been accompanied by decreasing energy efficiency but there are already indications that the rate of increase in crop yields is slowing down and that the point of diminishing returns may have been reached in cereal production (Fig. 14.5).

Intensification of agriculture has also incurred considerable environmental costs resulting from the high degree of management of its inorganic and organic compo-

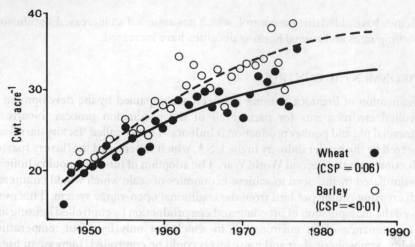

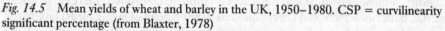

Fig. 14.5 Mean yields of wheat and barley in the UK, 1950–1980. CSP = curvilinearity significant percentage (from Blaxter, 1978)

nents. It has, on the one hand, created ecological problems, such as vulnerability to disease and soil deterioration, which not only increase production costs but put the stability of the agro-ecosystem at risk. On the other hand, intensification of agriculture has been accompanied by increase in waste products of both crop and livestock production, the disposal or loss of which create problems for other ecosystems on which they are deposited.

15

Agriculture and the environment

No agro-ecosystem can exist and function as an isolated self-contained self-sustaining unit. All are, by their very nature, more open than any unmanaged and than most other managed ecosystems. Even in the least technically developed type of subsistence agricultural system there has to be an input of energy in the form of human and/or animal labour; and, of the food produced and consumed, some of the waste in the form of human or animal excreta will be evacuated outside the immediately cultivated areas. In addition, disturbance of pre-existing 'natural' ecosystems will tend to increase the loss of nutrients by increased leaching, surface run-off and accelerated soil erosion. The relative intensity of agriculture in terms of number and volume of inputs and outputs has increased rapidly with the application of technical and scientific knowledge to agriculture since the 1940s. Intensity of agriculture also varies spatially, depending on the severity of the environmental constraints under which it operates and with the stage of regional development.

With increasing intensity of farming, interactions between agro-ecosystems and other non-agricultural systems have escalated. The outputs of one have become the essential or non-essential and even harmful inputs of another and vice versa. Further, the intensive production of crops and livestock has acquired many of the characteristic features of the manufacturing industry such as a high capital expenditure on machinery and buildings, specialization of production, increasing efficiency of production and a large output of waste byproducts. Agriculture waste material is now a significant source of air, water and soil pollution and an externality which imposes often high economic and social costs on other systems. Conversely, agro-ecosystems have become the recipients of pollutants produced elsewhere.

Until relatively recently concern about problems of increasing pollution and the associated environmental degradation focused attention mainly on urban–industrial complexes as the main sources of pollution. It was not until after the Second World War that the environmental impact of agriculture began to make itself felt. Awareness of the actual or potential consequences were heightened by the rate of change in agricultural methods at a time when public and private concern for conservation and environmental quality was becoming vocal.

ACCELERATED SOIL EROSION

The first and still the most deleterious impact of agriculture on the environment is accelerated soil erosion. Soil erosion is a natural process which has been speeded up as a result of the removal of a natural or semi-natural vegetation cover, and the exposure of the underlying soil to the direct impact of precipitation, sunlight and wind. Tillage disturbs the soil and opens it up. As a result, increased temperature and aeration speed up the decomposition of organic matter; and this is accompanied by a decline in the structural stability of the soil. It becomes more susceptible to transport by either wind or water, depending on local climate and land gradient. Accelerated soil erosion would have been initiated by deforestation and soil tillage. It early affected the highly populated and technically advanced agricultural societies of the Middle and Far East (particularly Northern China). The eventual decline of these societies has been attributed, at least in part, to the ravages of soil erosion and the decline in the efficiency of irrigation schemes and/or river flow as a result of the progressive silting of reservoirs and water channels. Awareness of the magnitude and universal extent of soil erosion emerged in the late 1920s and early 1930s, when cultivation was pushed into the semi-arid temperate grassland areas of the world. With the onset of a period of prolonged drought in the 1930s the exposed dry soils started to blow; and the impact of the Dust Bowl conditions in South-Western USA was felt as far afield as the eastern seaboard. In the period immediately prior to the Second World War, soil erosion was as emotive an issue as pollution or acid rain is at present. The publication of Jacks and Whyte's (1939) book *The Rape of the Earth* highlighted the global dimensions of the process.

Soil erosion continues to be the most virulent and widespread man-induced 'environmental disease' (Jacks, 1954). Despite the introduction of soil-conserving anti-erosion methods of cultivation, accelerated erosion continues wherever bare soil is exposed. Changes in land use and modern intensive methods of cultivation have served to exacerbate the problem, even in the more humid temperate areas of the world. Continuous cultivation, with an increase in the amount of land left bare in winter and a decline in soil organic-matter content, has resulted in the blowing of lighter drier soils and the sheet washing, rilling and gullying of heavier soils.

Precise data on the amount and rate of soil erosion are relatively sparse and are most frequently related to widely distributed local 'erosion events' or field experiments. Various estimates of the current net global soil loss have been made. Global erosion in 1970 has been put by Brown (1984) at 14×10^9 tonne yr^{-1} compared with 11×10^9 in prehuman times; and the same author has estimated that soil loss in the four major food-producing countries of the USA, USSR, China and India (which together account for half of the world's cropland) is 13.6 billion tonne yr^{-1}. In the USA, where so much experimental and applied work has been done on the nature of accelerated soil erosion and its control, it has been suggested that during the course of the last 200 years at least a third of the country's cropland top-soil has been lost (Pimentel *et al.*, 1976). Further, it has been calculated that the average loss of top-soil from agricultural land varies from 12 to 28 tonne ha^{-1}. The severity of erosion is dependent on a number of variables of which cultivation methods have played an increasingly important role. Row-cropping and continuous cultivation can result in an annual soil-erosion loss of up to 40 tonne ha^{-1}; in comparison, annual soil loss from grassland is only 0.6 tonne, and from forest land 0.002–0.004. It would appear that the effects of increased intensity of crop production have more than offset the reduction in soil erosion effected by the soil-conserving methods of cultivation initiated in the 1930s. Annual sediment loss by surface run-off from agricultural land in the USA increased from about 3 billion tonne in the 1930s to 4 billion in the 1970s (Pimentel *et al.*, 1976).

Accelerated soil erosion is currently a major environmental problem in the tropical and subtropical areas of the world as a consequence of population growth and increasing demands for food. Continuous or short-rotation cultivation has replaced longer fallows; and cultivation has, in many areas, been pushed further into areas where either the climate and/or soil and land forms are marginal for cultivation. Kowal and Kassam (1978) have suggested that the rate of accelerated soil erosion from cleared savanna land is now 20–200 times greater than the rate of soil erosion from undisturbed vegetation. The same authors note that soil loss can vary from 0.1 to 99 tonne ha^{-1} depending on environmental conditions and land management; under improved conditions, however, the annual loss from cropped land of 7–10 tonne ha^{-1} was comparable to that from cropland in the USA. In terms of actual acceptable limits of soil erosion (USA 0.4–1.8 tonne ha^{-1} yr^{-1}; East Africa 1.8 tonne for clay soils and 1.5 tonne for sandy soils) this is regarded as moderate (see Table 10.5b). Particularly high rates of soil erosion, as a consequence of increased rates of forest clearance for agriculture in the Himalayan country of Nepal, are now being recorded. In this instance severity of erosion is related to high rainfall erosivity combined with very long steep slopes and relatively easily erodible soil parent material.

Material lost by wind erosion from farmland can be transported considerable distances as evidenced by dust records from the African Sahel (see p. 206). In some cases the eroded material may be deposited on other farmland, where its impact will depend on its nutrient status and stability. Short-distance transport of sand particles (mostly SiO_2) can bury relatively fertile land under shifting sand dunes such as occurred on the margins of the American Dust Bowl in the 1930s and is currently taking place along the southern margins of the Sahel. Of the soil lost by water erosion, some will be deposited on the gentler valley and floodplain slopes. A very high proportion, however, finds its way into freshwater bodies and eventually into the oceans. The resulting sedimentation is one of the most prevalent and widespread forms of water pollution. It impairs water quality and reduces the efficiency of reservoirs, irrigation channels and navigable waterways as well as increasing liability to river flooding. In addition, soil erosion can result in nutrient losses from farmland in the USA of the order of 10 kg phosphorus, 10–20 kg nitrogen and 100–200 kg combined carbon ha^{-1} yr^{-1} (Pimentel *et al.*, 1976). It was further estimated that 6 kg ha^{-1} yr^{-1} (or 60 per cent) of applied phosphorus fertilizer was lost by soil erosion, while wind erosion of 10 mm ha^{-1} could remove 150 kg phosphorus from rich arable land. In North Africa, Kowal and Kassam (1978) note that an average annual soil loss by run-off of 200 mm contained approximately 22.7 kg ha^{-1} of the cations of sodium, potassium, calcium and magnesium and 13.7 kg ha^{-1} nitrogen. The eroded mineral soil retains a certain proportion of the less-soluble phosphorus and potassium, which are retained or fixed in an unavailable form by the soil clay minerals. The run-off from arable land then contributes to the nutrient enrichment or eutrophication of the waters into which it drains. The result is an increased rate of algae growth, increase in water turbidity and depletion of oxygen. Aerobic plants and animals die and the rate of decomposition and recycling of nutrients can decline to a level at which a freshwater lake may be regarded as virtually dead.

HABITAT CHANGE AND MODIFICATION

An inevitable consequence of the development and spread of agriculture has been the destruction of pre-existing wild or relatively unmanaged ecosystems, and the drastic modification of the physical environment by cultivation (see pp. 42–59). Habitat change or destruction can also be effected by land drainage designed either to convert wetland to agricultural use or to improve the soil water balance on farmland. In either case the impact of drainage can extend beyond the area of direct application and, in consequence, modify to a greater or lesser degree the soil water regime of adjacent non-agricultural ecosystems. Similarly the impact of domestic grazing animals and of vegetation burning in the interests of increased forage production can impinge on non-farm systems. This is exemplified in those situations where fire and/or grazing can inhibit tree growth and regeneration in natural or planted woodland or forests. In many highly and intensively cultivated parts of the world the continuing rapid reduction of remaining wild habitats by plough-up of rough/permanent grassland and heather moorland has given rise to concern about the implications for nature conservation. While some plant and animal species can adapt to, and have adapted to, the direct or indirect impact of particular agro-ecosystems, others have been unable to do so. Habitat reduction or destruction is inevitably accompanied by the decline and/or disappearance of plant and animal species. To date it is calculated (Lowe *et al.*, 1981) that Britain has lost:

95% herb-rich grassland
80% chalk/limestone grassland
60% lowland heath
45% limestone pavement
50% ancient woodland
50% fens and marshes
60% lowland raised bogs
33% upland grass and heather moor

This has been accompanied by the loss of animals whose existence was partially dependent on these particular habitats.

HEDGEROWS

The removal of hedgerows in the interests of field enlargement has attracted particular attention in the UK. The rate of removal has in recent years been so rapid in certain areas as to cause a very obvious change in the character of the landscape.

Between 1945 and 1970 it is estimated that approximately 1 per cent of the existing hedgerows were removed annually (Lowe *et al.*, 1981); reduction is concentrated primarily in the cereal-growing areas of the East of England, where some 50 per cent of the national loss of hedgerows has taken place on only six million acres (Caborn, 1971). Removal is, in fact, costly and the process has been encouraged to a certain extent by grant aid. The agricultural benefits have been an increase in the operating efficiency of large machines, together with a gain in workable land, as well as savings on hedge maintenance and on the wasted overlap of distributed seed, fertilizers and pesticides etc. The costs have been loss of shelter (less important in arable than stock-rearing farms) and, as a consequence, increased susceptibility of continuously cultivated soils to accelerated erosion. Caborn (1971) further notes that while a typical farm hedge can rarely provide much protection against soil erosion in large fields, the presence of hedges at recurring intervals may help to control localized 'foci' of wind blowing.

The most significant ecological impact of hedgerow removal has undoubtedly been the loss of associated plant and animal species and a reduction in the biological diversity of the agro-ecosystem. The hedgerow has been described as an impoverished woodland or woodland-edge habitat (Pollard *et al.*, 1974) in which a high percentage of small mammals, reptiles, birds, butterflies and insects breed. It has been estimated that 21 of the 28 species of mammals; 61 of the 91 birds; and 23 of the 54 butterflies in arable lowlands do so; and that removal would affect 20 of the total (250) associated species very seriously; 20 quite seriously and 20–30 to a marked extent. In addition, hedgerows have for a long period of time provided migration routes for wild animals and plant species; and a reduction in the density of this 'network' below a critical level is considered to have caused a serious reduction in bird populations. The species diversity of the hedgerow in comparison to the homogeneity of the adjacent crop community has both agricultural benefits and costs. On the one hand, the hedge harbours valuable game, such as partridge, and potential predators of agricultural pests. On the other, it is regarded as a source of weeds and pests of agricultural crops. On balance the agronomic value of the hedgerow depends on the particular farming system. The conservation value which is lost when the hedgerow is removed is a function of its dimension, form, age, condition, management and extent. In many remaining hedgerows the conservation value has been drastically reduced because of age and neglect in the face of high costs of management.

FARM WASTE MATERIALS

Among the agricultural impacts on the environment that have attracted much attention in recent decades are those resulting from the rapid escalation in the amount of chemical fertilizers, organic waste material, and pesticides.

CHEMICAL FERTILIZERS

Surplus soluble inorganic fertilizers – particularly nitrogen – which have not been taken up by the crop or fixed in the soil are leached out of the system. Others such as phosphorus and potassium are not so susceptible to loss by leaching, as is nitrogen, except under rather abnormal soil conditions. Surplus phosphorus is fixed in little-soluble inorganic or organic compounds and is only subject to significant loss in coarse sandy soils with little or no clay content. Most of the mineral potassium applied is also absorbed by the growing crops. The nutrient which 'leaks' from the soil in the greatest amounts is nitrogen, in the form primarily of nitrate or, to a lesser extent, ammonia.

ORGANIC WASTE

Agro-ecosystems produce varying amounts of waste material in the form of animal excreta and crop residues. The principal types of organic waste include:

Table 15.1 Approximate estimates of nitrogen involved annually in UK farming systems

	Nitrogen (tonne $\times 10^{-3}$)
Soil nitrogen (natural fixation, mineralization)	1300
Farmyard manure	300
Fertilizer	900
Excreta dropped on grass	500
Total	3000

(from MAFF, 1976d)

1. *Animal excreta* (urine and dung): while much of this may be returned to the farmland in the form of manure and/or slurry, a varying amount is lost from the system as a result of the deliberate spreading of slurry or as run-off from manure or animal droppings exposed on the soil surface. On large livestock farms application of slurry has increased in the face of the higher costs (time and labour) of transporting bulky manure.
2. *Silage effluent*: the amount produced is dependent on the efficiency of the process; poor silage can produce up to 228 litre tonne^{-1}.
3. *Cleaning effluent*: this results from the washing down of animals and the cleaning of animal houses and equipment. It is now produced in large amounts, particularly from intensive dairy farms, poultry units and piggeries.

All these organic wastes contain varying amounts of water, mineral nutrients, organic matter and other chemicals such as pesticides and other additives, which become incorporated in the animal feed (see Tables 15.1 and 15.2). They are evacuated in effluent form either by direct surface run-off; by agricultural drains; by soil interflow into rivers or lakes; or by percolation and seepage into ground water.

More recently the widespread development of intensive livestock farming in temperate areas has created new, very large and highly concentrated (i.e. 'point') sources of nitrate and phosphate pollution associated with changes in methods of

Table 15.2 Concentrations of nutrients in land drainage from arable and grassland areas at Saxmundham, 1973–74

| | Concentration ($mg\ l^{-1}$) | |
Nutrients	Arable	Grass
NO_3 (N)	22.0	4.0
NH_4 (N)	0.6	0.6
Cl	137.0	35.0
SO_4 (S)	81.0	54.0
Ca	215.0	108.0
Mg	9.0	7.0
K	1.2	0.9
Na	22.0	22.0
PO_4 (P)	0.02	0.01

(from MAFF, 1976d)

production, feeding, slaughtering, transporting and processing animal products. One of the most significant has been the development of 'housed' or 'confinement' feeding. An extreme form is the cattle feed-lot of South-Western USA, and the hen batteries and piggeries, which have become all but universal in distribution. These are large units holding from 1000 to 100 000 head of cattle at average densities of one animal to 5–15 m^2 (Leohr and Hart 1970) with a production of animal waste of *c*. 2500 tonne ha^{-1}; and up to 100 000 hens on sites of 2.5–3.5 acres. With little or no other farmland, disposal of waste has become an increasingly serious problem and a source of deleterious environmental impact. Under such climatic conditions

Table 15.3 Nitrogen and phosphorus contributed to natural waters in the USA

| | Nitrogen | | Phosphorus | |
	Quantity (tonne $\times 10^3\ yr^{-1}$)	Usual concn ($mg\ l^{-1}$)	Quantity (tonne $\times 10^3\ yr^{-1}$)	Usual concn ($mg\ l^{-1}$)
Domestic waste	454–725	18–20	91–227	3.5–9
Industrial waste	>454	0–10 000	–	–
Rural run-off				
agricultural land	680–6800	1–70	54–540	0.05–1.1
non-agricultural land	181–862	0.1–0.5	68–340	0.04–0.2
Animal waste	>454			
Urban run-off	50–500	1–10	5–77	0.1–1.5
Rainfall	14–268	0.1–2.0	1–4	0.01–0.03

(from MAFF, 1976d)

as heavy summer thunderstorm precipitation and rapid snow- and ice-melt in spring, surface run-off can cause direct and serious river pollution. Further, unlike other more diffuse sources of livestock waste, these new point sources (like human sewage) contain significant amounts of phosphorus as well as nitrates (see Table 15.3). In many parts of the USA the pollution equivalent derived from animal waste exceeds that from the human population by 5 : 1 in terms of the biological oxygen demand (BOD); 10 : 1 on a total dry solid basis and 7 : 1 on a total nitrogen basis. If only 10 per cent were to reach surface and ground waters, the pollution potential would, in fact, be equal to that of the human population.

However, only a small percentage of livestock waste is evacuated directly into surface water; this type of pollution is, in most developed countries, legally controlled. A higher proportion leaves the agro-ecosystem via the soil and, because of the relative immobility of phosphorus, nitrogen in the form of nitrates or ammonium is the main nutrient to be leached into surface and ground water.

NITRATE POLLUTION

Since the 1960s there has been mounting concern about the high and increasing levels of nitrate in river (Table 15.4) and ground water. In some areas of Britain these now exceed the recommended $11.3 \text{ mg } 1^{-1}$ set by the World Health Organization on the basis of health risks. Since 1977 several rivers in the Thames and Severn–Trent catchments in England have had mean annual concentrations of nitrogen in excess of $10 \text{ mg } 1^{-1}$; and in this area the assumption is that much of this comes from the intensive arable agriculture in this part of the country. Nitrates have been identified as one of the causes of the fatal, though rare, condition, methaemoglobinaemia in young babies. In addition, they are a source of carcinogenic nitroso compounds. Intensive arable agriculture is undoubtedly one of the main sources of nitrate pollution. Water draining from cropland can contain on average 10–$15 \text{ mg } 1^{-1}$ nitrate in a year's discharge (Dix, 1981) with concentrations considerably in excess of this level in winter, the period of maximum loss of the nutrient by leaching. The amount of nitrate lost from the land varies, however, dependent on the amount applied; the volume of water involved; the soil texture, which influences the rate of leaching and of denitrification; and the organic-matter content of the soil. In contrast, losses from grassland are much lower ($c. 5 \text{ mg } 1^{-1}$) except when heavy rain and saturated soil gives rise to surface run-off from densely stocked fields.

While nitrate levels have increased in rivers draining intensively farmed arable land this cannot, for a variety of reasons, be directly correlated with the quantities used. Nitrate concentrations remain highest in the winter months. The period of heaviest fertilizer use from March to May is, on average, only about a quarter of the total. In the latter period uptake by the growing plant and decrease in leaching consequent upon high evapotranspiration rates reduce losses. In the former, increase in nitrate levels can be associated with cessation of uptake and with nitrification leading to a build-up in the soil. This then becomes susceptible to leaching as the precipitation begins to exceed evapotranspiration in the winter. A high proportion of the intensive arable farming in the UK occurs in areas where ground water is a major source of domestic water and which now contains high levels

Table 15.4 Estimated breakdown of changes over time from 5 year averages – all figures quoted as $kgNha^{-1}$

STOUR: (Catchment Area = $578\,km^2$)

			Changes over time of individual inputs		
Years	Annual load* from all sources	Sewage component	Rainfall	Fertilizer usage	Ploughing of grass
1940–4	5.46	1.73	1.79	15.09	4.76
1945–9	4.62	1.73	1.84	18.99	0.76
1950–4	7.96	1.73	1.91	22.28	0.05
1955–9	5.88	1.73	1.82	31.35	0.36
1960–4	7.48	1.79	1.81	45.34	1.26
1965–9	13.52	2.06	1.84	57.46	3.30
1970–4	12.61	2.33	1.56	62.98	0.50
1975–9	17.40	2.66	1.78	74.82	0.50

(* Total NO_3 as $kgNha^{-1}$ calculated from river flow × nitrate concentration i.e. not sum of potential inputs)

THAMES: (1975 figure) (Catchment Area = $9950\,km^2$)

			Changes over time of individual inputs		
Years	Annual load* from all sources	Sewage component	Rainfall	Fertilizer usage	Ploughing of grass
1930–4	5.87	0.86	2.02	3.90	insignificant
1935–9	7.49	0.90	2.35	4.90	
1940–4	6.46	0.93	2.00	11.62	4.73
1945–9	6.81	1.00	2.08	14.62	insignificant
1950–4	10.59	1.07	2.34	16.90	
1955–9	9.39	1.17	2.03	22.52	0.38
1960–4	10.05	1.27	2.12	39.42	2.84
1965–9	12.28	1.39	2.31	54.70	3.78
1970–4	13.35	1.51	2.09	72.90	0.70

(* Not sum of individual inputs)

(from Department of the Environment, 1984)

of nitrates. Pollution risks are highest where annual leaching is of the order of 30–50 $kg\,ha^{-1}\,yr^{-1}$. Further, Dutch workers consider that where livestock densities are of three cattle or more units per hectare, overmanuring can lead to pollution of surface waters (MAFF, 1976d). However, the passage of polluted water downwards from the soil would, normally, unless facilitated by rock joints and fissures, tend to be slow. It is thought that it could take up to 30 years or more before the full impact of increased nitrate pollution is reflected in river water. Average nitrate levels now exceed the recently imposed EEC limits of 50 mg 1^{-1} in approximately a fifth of the ground-water sources in East Anglia.

The report of the symposium on *Agriculture and Water Quality* (MAFF, 1976d) concluded that more of the nitrogen leakage from farm systems came from the mineralization of soil organic matter – a natural process speeded up by cultivation – than from surplus nitrogen fertilizers, which probably contribute no more than *c*. 5 per cent of the total load. The *Agriculture and Pollution* (MAFF, 1979) stated that, while agriculture is a major source of nitrate pollution, it was then not yet a major health hazard on its own. What is more significant is that many agricultural water-catchment areas cannot now supply water sufficiently low in nitrates to be used effectively in the dilution of urban sewage and industrial effluent.

PESTICIDES (see Table 15.5)

A considerable number of substances have been used, at one time or another, to control agricultural pests – the unwanted plants, animals and micro-organisms. It was not, however, until 1850 that the term 'pesticide' came into general use in agriculture. Before the Second World War the most important pesticides were the relatively 'safe' natural organic derivatives such as nicotine, pyrethrum and roten-tone together with lesser amounts of Bordeaux mixture ($CuSO_4$), mercury, zinc and sulphur. The amounts used, however, were small and irregularly applied. Since 1940, these early pesticides have been superseded by a large array of complex synthetic substances which are more effective in controlling pests but which can have side-effects of varying severity on wild plants and animals within and beyond the farmland on which they are applied.

Many of the first-produced synthetic pesticides were 'systemic' in action, capable of penetrating plant tissues. They were more effective than the earlier 'contact' or 'surface' substances, which were susceptible to weathering and therefore were more variable in action (Dix, 1981). The systemic pesticides not only killed the pests on the plants but also those in the plant tissues. The development of pesticides to control animal pests – the *insecticides* – was followed in the 1950s and 1960s by the production of ever-more specific compounds, e.g. *fungicides* and *herbicides* etc. The early recognition that some of a pest's population were resistant to a given pesticide, and that resistance increased over time, stimulated the production of ever-more effective chemicals for particular groups or for a particular species of pest. As a result, the number of substances used has continued to multiply. By the mid-1970s it was estimated that over 900 different chemicals were used as pesticides in the USA; the majority, however, could be classified into twenty-five chemical types (Crosby, 1973). The active substances are produced for use in a variety of forms – as dusts, wettable powders, emulsifiable concentrates etc. – which can have an import-ant influence on the dispersal and breakdown of the particular pesticides.

Normally the area occupied by a crop represents only a relatively small propor-tion of the pest's habitat range. Consequently spring infestation potentials (Strick-land, 1960) are so high that under optimum conditions for dispersal a pest may be

Table 15.5   Some pesticides in common use

Insecticides

Natural organics	nicotine	
	rotenoids (derris)	
	pyrethrum	

Synthesized organics

organochlorine	DDT	aldrin
	gamma-BHC (Lindane)	dieldrin
	chlordane	endrin
organophosphorus	TEPP	dimethoate
	metasystox	menazon
	malathion	chloropyrifos
	dichlorvos ('Vapona')*	phorate
carbamates	carbaryl ('Sevin')	propoxur

Fungicides

Surface types	phenyl/alkyl mercury	captan
	thiram	dinocap ('Karathane')
	zineb	quintozene
	maneb	rovral
Systemic types	benomyl	triarimol
	thiophante-methyl	'Dowco'
	carboxin	'Milstem'

Herbicides

Non-selective	simazine	diquat
	glyphosate	paraquat
	monuron	
Selective growth regulators	MCPA or B†	mecoprop
	2,4-D†	dichloroprop
	2,4,5-T*†	
	maleic hydrazide	dalapon†
	chloromequat	TCA

* Quotation marks indicate trade mark names.
† These herbicides can be used non-selectively during the growing season.

(from Dix, 1981)

able, even in a year of minimal population, to swamp the cropland many times over, as is illustrated in Table 15.6. In consequence there has been a tendency to use larger amounts and higher concentrations of pesticide to cope with the problems of resistance and spring infestations. It has been estimated that in 1977 55 per cent of British farmers sprayed cereal crops and 43 per cent sprayed other arable crops once to three times per season, while 42 per cent and 46 per cent respectively sprayed the same crops four to six times (Dix, 1981). It has, however, been shown that application efficiency is often low, with less than 20 per cent of the material applied

Table 15.6 Ecological problems in crop pest control: spring infestation potentials for two pests

	Beet leaf-hopper	Fruit fly
Zonal habitat	S. Idaho and E. Oregon	England and Wales
Total productive land*	13	28
Area in which wild hosts are very important	3.9	11.0
Area of rough grazing of some importance to the pests	7.8	3.6
Area of susceptible crops	0.2	1.2
Estimated total insects emerging in the spring ($\times 10^9$)	70–850	30–160
Estimated numbers of insects economically tolerable on the host crops (10^3 per acre)	6–11	3–6
Spring infestation potential therefore sufficient to infest million acres of crop	12–142	15–53

* All land areas are given in acres $\times 10^6$.

(from Strickland, 1960)

reaching its target. In a particular application study quoted by Crosby (1973) the actual efficiency of methoxychlor to alfalfa in California was less than 0.05 per cent.

Pesticide residues are disposable into the air, soil and water, and other non-agricultural ecosystems, where they are subjected to biological and non-biological transformation. Most pesticides enter the agro-ecosystem via the green plants either by direct foliar absorption or by uptake from water or soil. Some of that in the soil is applied directly; some is incorporated in dressed seed. Up to 50 per cent of the foliar applications can be washed off the leaves into the soil, while significant quantities of pesticides can be transported for great distances as dust particles in the atmosphere, later to be carried into the soil by rainfall. In addition, plant and animal residues return pesticides to the soil. 'The soil at some time forms a vast repository of much of the world's pesticides' (Crosby, 1973: 481). Of the organic pesticides, the organochlorines are the most generally distributed. It has, for instance, been calculated that in areas of prolonged and heavy application of DDT hundreds of kilograms per hectare might accumulate locally, representing most of the pesticide ever applied.

While a considerable quantity of pesticide is added to fresh water to control weeds, algae and insect larvae (particularly mosquitoes), larger amounts are introduced by drift, disposal and by domestic and industrial sewage. The atmosphere also receives an input of pesticide as a result of spraying. Dust and particulate matter can carry pesticides over great distances: detectable levels of persistent compounds such as DDT have been recorded from Antarctic ice and remote oceanic islands. Local concentrations can be very high at the time of application.

The effect of pesticides on other than the target species varies, depending on their toxicity and persistence in the environment. The latter is a function of the stability of the particular chemical, i.e. its resistance to biological or non-biological degradation. Crosby (1973) states that the majority of pesticides are not particularly persistent and there are no permanent organic pesticides. The rate of degradation into simple inorganic elements depends on a complex of interacting environmental variables including climatic conditions, soil type, plant cover etc. The organo-chlorine group are highly toxic and are among the most stable and persistent of the insecticides. Initial concern about the impact of pesticides on the environment was stimulated by the publication of Rachel Carson's (1963) book *Silent Spring*, which drew attention to the actual and potential threat of these new chemicals to wildlife – particularly to birds. Wildlife, fish and domestic animals including man were shown to contain varying amounts of DDT and other chlorinated hydrocarbons. The only significant source for most animals was via the plant food consumed, except in the case of fish, which can absorb these substances direct from water. Particularly high and often toxic insecticide levels (over 10^4 times that in the environment) in fish was interpreted as evidence of biological concentration or magnification.

If insecticides not voided from the animal system are stored in certain organs, such as the liver, they will be ingested by the next predator. The absolute amount will remain the same but the concentration will increase consequent upon the loss of energy in conversion from one trophic level to another. This being the case the animals at highest risk are those such as predatory birds and fish at the end of food-chains. On the other hand, high insecticide levels can result from inefficient excretion and the accumulation of substances faster than they are voided and returned to the environment; and continuous intake would be accompanied by temporary storage in the animal tissues. By whatever route the organochlorines enter the system they are highly toxic at relatively low levels; and there is little doubt that they have contributed to the marked decline or elimination of certain particularly vulnerable species of wildlife. In some cases this is as a result of death by direct poisoning and in others of a decline in reproductive capacity as evidenced by thin-shelled and improperly formed birds' eggs. A comparative survey of butterfly numbers on sprayed and unsprayed arable farmland in South-East England suggested that pesticide use on cereal crops may be the main factor in the reduced numbers on the sprayed land (Rands and Sotherton, 1986).

At present the USA and Western Europe account for about 70 per cent of all pesticides used. With increasing intensity of agriculture and continuous cultivation the amount of pesticide applied has increased steadily. The banning of the use of those such as DDT and its derivatives and paraquat (which is particularly lethal to handle) has been accompanied by the continuing search for alternatives which are as, if not more, effective. In addition, the need for and use of pesticides in the developing tropical countries of the world is starting to grow. While it would probably be true to contend that, as in the temperate regions, the environmental effect of pesticides will be conditioned by their persistence and accumulation in organic tissues and in the environment, it is not possible to assume that a particular type of pesticide will necessarily behave in the same way in the tropical as in the temperate environment.

ENVIRONMENTAL IMPACT ON AGRO-ECOSYSTEMS

As well as providing inputs into other ecosystems, agro-ecosystems have themselves been increasingly subjected to inputs from the surrounding non-agricultural environment. The most pervasive of these is the pollution of air and soil. The main sources of air pollutants such as dust (i.e. particulate matter), sulphur dioxide (SO_2), acid rain and nuclear fall-out tend to be concentrated in or near large urban-industrial areas, though many of the more recently constructed large thermal and nuclear power stations and petrochemical complexes are being located in the countryside.

While the potential impact of air pollution on vegetation in general is known, the effect on agricultural production has always been assumed to be relatively small because of distance from pollution sources, though it has never been precisely quantified. However, in particular isolated cases, air and/or soil pollution have been suspected, though not definitively proved, of being the cause of serious agricultural damage. These include fluorosis and reduction in yield of vines in Australia (New South Wales), which have been attributed to the production of fluoride by large aluminium smelters located in the vicinity. In addition, dramatic incidents such as the death of large numbers of sheep in the USA and cattle in the UK have been associated either with known pollution events or with suspected but non-proven soil and air pollution from nearby chemical plants. However, the most dangerous impact has been that resulting from nuclear fall-out, the result of accidents such as those at Three Mile Island (USA), Windscale (UK) and the most recent and extensive at Chernobyl (USSR) in spring 1985.

Radioactive isotopes in nuclear fall-out include strontium-90, caesium-137 and iodine-131, among others. The first two are long-lived and decay much more slowly than the last. They can be absorbed directly from air, water and soil by plants and, if not degraded in the process of respiration or excreted, they can become progressively concentrated in animal and human tissues (e.g. strontium in bone; caesium in flesh) as they pass from lower to higher trophic levels in the agro-ecosystem. Dairy and upland stock farms and reindeer ranches in the high rainfall areas of North-West Europe were most seriously and directly affected by fall-out from Chernobyl: in this case caesium-137 was taken up by the grazing animals to an extent that their becquerel levels increased beyond that regarded as acceptable for human consumption. In some cases embargoes were put on the slaughter and sale particularly of hill sheep in South-West Scotland, Cumbria and Wales until, with growth, it was assumed that concentrations had reached a lower level. A recent newspaper report (*The Guardian*, 21 January 1988) suggested that the implications for sheep farmers in North Wales (where sheep grazing on peatland contaminated by local sources as well as by Chernobyl are giving readings of 3000 Bq) could last for 30 years because of a change in acceptable levels from 1000 to 600 Bq. In others, as in the case of reindeer in Scandinavia, herds were slaughtered in the interests of long-term genetic safety.

More direct and quantifiable are the depredations of protected wildlife species on crops. Conservation of rare or endangered species has in some instances been so successful that populations have recovered and grown to become damaging pests of agricultural land, as in the case of the red and roe deer and wildfowl such as Brent geese in Britain.

Finally, the competition for and sterilization of first-class agricultural land by urban, industrial and transport developments is made the more serious by the relative paucity of this type of land in Britain and other European countries. It has generated considerable concern in the USA, where the *National Agricultural Land Study* (USDA, 1981) was set up to examine the nature, rate, extent, cause and the long-term consequences of the conversion of land from agricultural to non-agricultural uses. In all the major developed countries the rapid growth of long-distance multilane motorways and pipeline constructions for water and natural gas has cut deep into or under agricultural land. Loss of land has often been exacerbated by severance of holdings and disruption of agricultural drainage systems and of soil water regimes. Further, disturbed land along pipeline routes and motorway verges in Britain has become a new source of a group of agricultural weeds dominated by the *Umbelliferae* (wild parsley, hogweed etc.) and the poisonous ragwort. Rapid road-transit has also facilitated the transmission of highly infectious diseases of crops and animals, such as foot and mouth disease and swine fever, for which the only remedy is herd destruction.

The development of the motorway together with the tremendous increase after the Second World War in the numbers of people participating in outdoor recreation saw the growth in the perception of the countryside as the urban dwellers 'play-ground'. There is little doubt that in certain areas visitor impact on crops and animals is comparatively high whether it be destruction of crops, damage to farm structures, livestock 'worrying' by dogs, failure to shut gates or the increase in synthetic (plastic) litter. It is at its most intense at the urban–rural interface. Here farmland has either succumbed to the demand for urban development or changed over to types of agricultural or semi-agricultural use such as potato growing, market gardening or 'pony farms', all of which are more compatible with and resistant to urban pressures.

AGRICULTURAL IMPACT ON THE LANDSCAPE

Since the 1940s, concern for the conservation of wildlife in the face of a variety of land-use developments, including agriculture, has been closely associated with that for the aesthetic value of the countryside; and both have been stimulated by environmental education and the media's increasing interest in ecological problems on the one hand and by the increased pace of change in the visual appearance of the rural landscape on the other.

The countryside today is the product of a very long period of ever-changing types and methods of human use of which the most important has always been agriculture. Changes in the appearance of the landscape have taken place in the past (such as the enclosure of land by hedges, stone walls or fences), and until recently, relatively slowly as in the case of the replacement of the haystack by the silo. With increasing intensity, particularly of arable farming in Britain and the countries of North-West Europe, land management practices which were known or were assumed to damage the wildlife and/or aesthetic value of the countryside came under increasing criticism. In particular, the removal of field boundaries, enlargement of field size and the high degree of specialization in continuous cereal cultivation in the traditional arable areas of the lowlands of Eastern England produced what became characterized as a 'prairie landscape' distinguished by a featureless monotony of appearance. Other types of 'damage' identified and deplored included drainage schemes which effected a rapid disappearance of species-rich wetlands; the loss of heather-dominated moorland, particularly in South-West England in the face of the establishment of improved grassland; the destruction of or damage to ancient deciduous woodland; and even the construction of new farm roads in scenically attractive areas. While the impact of such practices, 'many of which would have been considered, until recently, as indicators of good husbandry and as a sign of an economic and rational use of resources' (*Agriculture and the Environment* (House of Lords, 1984)), was more in evidence in the small-scale landscape of Britain, they were also taking place in the arable areas of North-West Europe, where they have engendered the same concern for wildlife and landscape conservation.

Unfortunately there has been a tendency to equate all intensive or even moderately intensive farming with that of the large-scale cereal production. On company owned and/or managed holdings the emphasis, engendered by high prices and grant aid, has been and continues to be on high inputs to achieve high yields and to assume that all farmers are using their resources and 'mining' their soil in a similar way. Although government-generated incentives for high production have been considerable, costs of production have also been increasing at a greater rate than prices. Not only is the tradition of good husbandry maintained in many arable and livestock farms beyond the cores of high intensity, dependence on the input of inorganic fertilizers is not so high and ley grass, chopped straw and muck-spreading have maintained soil organic matter at higher levels than in the so-called 'prairie lands'. Greater crop diversity and the maintenance of some livestock have not necessitated the removal of field boundaries to the same extent as in Eastern England. The development of modern intensive agriculture has, however, brought nature conservation and farming interests into sharp conflict not only in Britain but elsewhere in Europe and North America. There has been, as a result, at times extreme polarization of their respective perceptions of each other's aims. More recently, however, there have been concerted and frequently very successful attempts to achieve a compromise between farming and nature conservation. This has resulted from a greater appreciation of the need for and aims of wildlife and landscape conservation coupled with problems arising from agricultural overproduction and the increasing public demand for food produced without the use of inorganic fertilizers, i.e. by what is now called *organic farming* (Oelhaf, 1978). In many areas of Britain there has been a revival of the grass/legume 'break' crop to help build up soil fertility and thereby reduce expenditure on increasingly costly inorganic fertilizers. Farmers are becoming involved in plans to conserve the remaining wildlife habitats on their land, as well as to protect these as far as possible from the direct impact of insecticides. Further, suggestions have been made that farmers should be given financial incentives to maintain traditional elements of the rural landscape such as field boundaries, woodlands, buildings etc.

Ways and means of reducing agricultural overproduction in Britain and other countries of the European Community arc being investigated and implemented. Financial incentives (comparable to those given for the same purpose in the USA in the 1960s) to take land out of arable cultivation and to put it to some other non-farm but income-generating use such as woodland, pony farming or some other kind of recreational/tourist use are being proposed. In the UK farmers are to be financially compensated for land taken out of cultivation, i.e. 'set aside'. While the scenic impact of intensive farming has been deplored, the long-term impact of agricultural withdrawal on the landscape and the extent to which the resulting changes would be more or less aesthetically and/or ecologically acceptable than the *status quo* have, in comparison, been given little serious discussion. Financial incentive alone, to withdraw land from agriculture, could merely encourage withdrawal from marginal areas and make for greater concentration of specialization and increasing productivity on land in the highest capability classes. Indeed, in upland areas throughout Western Europe, economic pressures are maintaining the steady retreat of agriculture; and land values for non-agricultural and even 'unproductive' use for private sporting, recreational or long-term speculative investment purposes have been increasing.

Land abandoned in the past has tended either to 'run down' and be accompanied by the rapid re-invasion of scrub species such as bramble, hawthorn, gorse; vigorous weeds, particularly bracken; and rushes and other plants associated with poor drainage; or to be taken over for the establishment of stands of commercial conifers. Field boundaries, fences, hedges, stone walls and farm woodland are neglected and the condition of agricultural buildings deteriorates. For example the breakdown of the traditional crofting system of farming in the Outer Isles of Scotland was accompanied by just such landscape changes, while the recent implementation of Integrated Developments Programmes, aimed at reviving the agricultural economy, is already bringing about the emergence of a rural landscape similar to much of lowland Scotland. Further, there are indications that the extensive land areas now marginal for agriculture may change in appearance even more rapidly and drastically than did the most productive land. Intensification of agriculture in the Mediterranean area of Western Europe has been accompanied by the decline of traditional farming systems based on the integration of cereal, olive and vine cultivation and sheep grazings on rough grazings of the adjacent garrigues. Demand for *vin ordinaire*, fruit and fresh vegetables has resulted in the development of a high degree of specialization with extensive monocultures of vines and citrus fruits and in high-intensity irrigation agriculture. The last has resulted in the decimation of former coastal wetlands. Irrigated farming now dominates areas of level land and heavier soils where there is access to ground water. These, however, are relatively small 'green oases' set in the more extensive 'drylands' which dominate the Mediterranean landscape and over which the scenic imprint of a former type of farming still persists in the form of crumbling stone walls and deteriorating farm buildings, intricate terraces climbing high up steep slopes and still supporting the long-lived olive tree.

In 1985 an ambitious massive aid scheme – the Integrated Mediterranean Programmes (IMPs) – was initiated by the European Community to give finance to twenty-nine regions in France, Italy and Greece. The aim is to phase out marginal agriculture and to improve the economic infrastructure of these rural regions and upgrade the physical environment. Industry including fisheries is to be promoted and alternative uses of marginal farmland, such as re-afforestation and recreational and tourist facilities are to be encouraged. Concern has already been generated about the consequent impact of the IMPs on one of the oldest rural landscapes in Europe – the product of a centuries-old system of farming. It has been suggested that it could seriously detract from the tourist, scenic and wildlife values of a traditional landscape about which Northern European perceptions have been maintained and reinforced by writers, painters and modern tourist agencies. Given the average holiday-maker's demands and aspirations this may seem an extreme point of view. It does, however, serve to emphasize the fact that methods and aims of agriculture have changed in the past and will do so in the future; and the maintenance of a 'constant' or relatively little-modified rural landscape requires the preservation and protection of the agricultural *status quo*. This can be and has been successfully done, at a cost, in 'protected traditional rural landscapes' in the Netherlands, Denmark and France. If agricultural, landscape and wildlife values are to be optimized, management to achieve this end in the face of changes in economic climate is needed. The *National Agricultural Land Study* (USDA, 1981) in the USA comes to the conclusion that conversion of agricultural land (in this case to urban or urban-related uses) does not present an immediate crisis. It does, however, hold some long-term risks given what the report regards as a progressive demand for agricultural products from outside the USA and because of the inevitable uncertainties about increasing crop yields per hectare at home. Hence it recommends that the Federal Government make the protection of good agricultural land a national policy. There are those who would echo this sentiment in Europe and point out that the long-term aims should be directed not towards how to reduce food surpluses by alternative land uses but towards maintaining the agricultural value of the soil resource at a *sustainable* level and as compatible as possible with the conservation of wildlife and of the scenic quality of the rural landscape.

References

Agricultural Research Council 1967 *The Effect of Air Pollution on Plants and Soil*, HMSO

Agricultural Research Council 1980 *The Nutrient Requirements of Farm Livestock* No. 2 *Ruminants*, HMSO

Ahmad, N 1979 Tropical Clay Soils – Their Use and Management, *Outlook on Agriculture* 13(2), 87–96

Aina, P O, Lal, R, Taylor G S 1979 Effects of vegetal cover on an Alfisol, in: Lal, R, Greenland D J (eds), 501–67

Alberda, T W 1962 Actual and Potential Production of Agricultural Crops, *Netherlands Journal of Agricultural Science* 10, 325–33

Alexander, M 1974 Environmental Consequences of Rapidly Rising Food Production, *Agro-Ecosystems* 1, 249–64

Allen, E J, Wareing, J 1977 Physiological Aspects of Crop Choice *Philosophical Transactions of the Royal Society* London B, 281–91

Allison, F E 1973 *Soil Organic Matter and its Role in Crop Production*, Elsevier, Amsterdam

Alvim, P de T, Kozlowski, T T (Eds) 1979 *Ecophysiology of Tropical Crops*, Academic Press

American Society of Agronomy 1976 *Multiple Cropping*, Special Publication No. 27, Madison, Wisconsin

Amin, M A 1977 Problems and Effects of Schistosomasis in Irrigation Schemes in the Sudan, in: Worthington, E B (Ed.), 407–12

Andrews, D J, Kassam, A H 1976 The Importance of Multiple Cropping in Increasing World Food Supplies, in: American Society of Agronomy, 1–10

Archer, J 1985 *Crop Nutrition and Fertiliser Use*, Farming Press

Atterberg 1911 Plastizität der Tone. *Intern. Mitt. Bodenk* 1, 10

Aubert, G 1974 *A World Assessment of Soil Degradation*, FAO, Rome

Austin, R B 1978 Actual and Potential Yields of Wheat and Barley in the United Kingdom, *ADAS Quarterly Review* 29, 76–87

Aweto, A O 1981 Total Nitrogen Status of Soils Under Fallow in the Forest Zone of South West Nigeria, *Journal of Soil Science* 32(4), 639–42

Ayanaba, A 1977 *Biological Nitrogen Fixation in Farming Systems of the Tropics*, Wiley

Bach, W, Pankrath, J, Schneider, S 1981 *Food-Climate Interactions*, Reidel Publishing Company

Bailey, H P 1979 Semi-arid Climates: Their Definition and Distribution, in: Hall A E *et al*, 73–97

Balch, C C, Reid, J T 1976 The efficiency of conversion of feed energy and protein into animal products, in: Duckham, A N *et al*, 171–98

Balls, W L 1953 *The yields of a crop*, E & F Spon

Barley, K P 1961 The Abundance of Earthworms in Agricultural Land and their Possible Significance in Agriculture, *Advancements in Agriculture* 13, 249–68

Barke, Mike, O'Hare, Greg 1986 *The Third World*, Oliver & Boyd

Batch, J S 1979 Recent Developments in Growth Regulators for Cereal Crops, *Outlook on Agriculture* 10(8), 371–8

Bayliss-Smith, T P 1982 *The Ecology of Agricultural Systems*, Cambridge University Press

Beek, K J, Bennema, J 1972 *Land Evaluation for Agricultural Land Use Planning: an Ecological Methodology*, Agricultural University of Wageningen, Netherlands

Beek, K J 1977 The Selection of Soil Properties and Land Qualities Relevant to Specific Land Uses in Developing Countries, in: *Soil Resources Inventories* No. 7, 7–23, Cornell University, New York

Beek, K J 1978 *Land Evaluation for Agricultural Development*, Institute for Land Reclamation and Improvement Publication 23, Wageningen, Netherlands

Beetz, W C 1982 *Multiple Cropping and Tropical Farming Systems*, Westview, Boulder, Colorado

Begg, J E, Turner, N C 1976 Crop Water Deficits, *Advancements in Agriculture* 28, 161–207

Bertin, J, Hemardinquer, J J, Kewl, M, Randles, W G L (Eds) 1971 *Atlas of Food Crops*, Ecole Practique de Haute Etudes, Mouton et Co, France

Bianca, W 1961 Heat Tolerance in Cattle, *Journal of Biometeorology* 1, 5–30

Bianca, W 1976 The Significance of Meteorology in Animal Production, *Journal of Biometeorology* 210, 139–56

Bibby, J S (Ed.) 1982 *Land Capability Classification for Agriculture*, Macaulay Institute for Soil Research, The Soil Survey of Scotland

Bibby, J S, Mackney, D 1969 *Land Use Capability Classification*, Technical Monograph 1, Soil Survey Great Britain, Rothamstead

Bingham, J 1961 The Achievements of Conventional Breeding, *Philosophical Transactions of the Royal Society London* B 292, 441–54

Biscoe, P V, Gallagher, J N 1977 Weather, Dry Matter Production and Yield, in: Landsberg and Cutting (Eds), 75–100

Biscoe, P V, Gallagher, J N 1978 A Physiological Analysis of Cereal Yield – Production of Dry Matter, *Agricultural Progress* 53, 34–58

Biscoe, P V, Scott, R K, Monteith, J L 1975 Barley and its Environment, *Journal of Applied Ecology* 12, 269–93

Biswas M R, Biswas, A K 1980 *Desertification*, Pergamon

Blacksell, M, Gilg, A 1981 *Countryside Planning and Change*, Allen and Unwin

Blake, P W (Ed.) 1985 *Livestock Production*, Heinemann

Blaxter, K L 1978 Limits to Animal and Crop Production, *The Journal of Australian Institute of Agricultural Science* 44, 97–103

Blaxter, K L 1986 *People, Food and Resources*, Cambridge University Press

Blaxter, K L, Fowden, L 1982 *Food Nutrition and Climate*, Applied Science Publications

Blaxter, K L, Fowden, L (Eds) 1985 *Technology in the 1990s: Agriculture and Food*, The Royal Society, Cambridge University Press

Blount, W P (Ed.) 1968 *Intensive Livestock Farming*, Heinemann

Bonin, D 1978 Cattle, Rainfall and Tsetse in Africa *Journal of Arid Environments* 1, 49–61

Bonner, J 1962 The Upper Limit of Crop Yield, *Science* 137, 11–15

Borchert, J R 1971 The Dust Bowl in the 1970s, *Annals of the Association of American Geography* 61, 1–22

Bowers, J K, Chesire, P C 1983 *Agriculture, the Countryside and Land Use*, Methuen

Bowman, J C 1977 *Animals for Man*, The Institute of Biological Studies: Studies in Biology No. 78, Arnold

Boyko, Hugo 1968 Farming the desert *Science Journal*, May, 72–78

Bradfield, R 1974 Intensive Multiple Cropping, *Tropical Agriculture* 51(2), 91–3

Brady, N C 1974 *The Nature and Properties of Soils*, 8th ed., Macmillan, NY

Bremner, H, De Witt, C T 1983 Rangeland Productivity and Exploitation in the Sahel, *Science* 221, 4618

Bresler, E, McNeal, B L, Carter, D L 1982 *Saline and Sodic Soils*, Springer-Verlag, Berlin

Briggs, D, Courtney, F M 1982 *Agriculture and the Environment*, Longman

British Society of Soil Science 1974 *Symposium on Land Capability, Edinburgh*, Collected Papers, Rothamstead

Brown, G 1974 The Agricultural Significance of Clays, in: Mackney, D (Ed.) *Soil Type and Land Capability*, Technical Monograph 4, Soil Survey, Harpenden 27–42

Brown, G D, Williams, O B 1970 Geographical Distribution of the Productivity of Sheep in Australia, *Journal of the Australian Institute of Agricultural Science* 36, 182–9

Brown, L R 1981 World Population Growth, Soil Erosion and Food Scarcity, *Science* 214, 995–1002

Brown, L R 1984 Global Loss of Topsoil, *Journal of Soil and Water Conservation* 39, 162–5

Buckman, H O, Brady, N C 1960 *The Nature and Properties of Soils*, 6th ed., Macmillan, NY

Bullen, E R 1975 How Much Cultivation? *Proceedings of the Transactions of the Royal Society London* B281, 153–76

Bunting, A H 1975 Time, Phenology and the Yield of Crops, *Weather* 30, 312–25

Buol, S W, Conto, W 1981 Soil Fertility Capability Assessment for Use in the Humid Tropics, in: Greenland, D J (Ed.) *Characterisation of Soils in Relation to their Classification and Management for Crop Production*, Oxford University Press, 254–61

Buresh, R J, Casselman, M E, Patrick, W H (Jnr) 1980 Nitrogen Fixation in Flooded Soil Systems: A Review *Advances in Agronomy* 33, 150–87

Butterworth, M H, Lambourne, L J 1987 Productivity improvement and resource conservation in the use of African rangelands, in: Joss *et al*, 111–4

Butterworth, B, Davidson, J G, Sturgess, I M, Wiseman, A J A 1980 *Arable Management*, Northwood Books

Butzer, K W 1974 Accelerated Soil Erosion – A Problem of Man Land Relationships, in: *Perspectives on Environment*, American Association of Geographers, Washington

Caborn, J M 1971 The Agronomic and Biological Significance of Hedgerows, *Outlook on Agriculture* 6, 279–84

Caldwell, M M 1975 Primary production of grazing lands, in: Cooper, J E (Ed.), 41–75

Cannell, R Q, Finney, J R 1973 Effects of Direct Drilling and Reduced Cultivation on Soil Conditions for Root Growth, *Outlook on Agriculture* 44, 184–9

Carlson, Peter S (Ed.) 1980 *The Biology of Crop Production*, Academic Press

Carson, R 1963 *Silent Spring*, Hamish Hamilton

Carter, D L 1975 Problems of Salinity in Agriculture, *Ecological Studies* 15, 26–35

Charney, J, Stone, P H, Quirk, W J 1975 Drought in the Sahara: A Biogeographical Feedback Mechanism, *Science* 187, 343–5

Chancellor, R S 1982 Dormancy in Weed Seeds, *Outlook on Agriculture* 11(2), 87–93

Chang, J-H 1968a *Climate and Agriculture: an Ecological Survey*, Aldine Publishing Company, Chicago

Chang, J-H 1968b The Agricultural Potential of the Humid Tropics, *Geographical Review* 58, 333–61

Chang, J-H 1970 Potential Photosynthesis and Crop Production, *Annals of the Association of American Geographers* 60(1), 92–101

Chang, J-H 1977 Tropical Agriculture: Crop Diversity and Crop Yield, *Economic Geography* 53, 241–54

Chapman, S R, Carter, L P 1976 *Crop Production: Principles and Practices*, W H Freeman, San Francisco

Charlton, C 1987 Problems and Prospects for Sustainable Agriculture Systems in the Humid Tropics, *Applied Geography* 7(2), 153–74

Clements, R O 1980 Pests: the Unseen Enemy, *Outlook on Agriculture* 10, 219–23

Clutton-Brock, J 1980 *Domesticated Animals from Early Times*, British Museum with Heinemann

Colbourn, P 1985 Nitrogen Losses from the Field: Denitrification and Leaching in Intensive Winter Cereal Production in Relation to Tillage Methods of a Clay Soil, *Soil Use and Management* 1(4), 117–24

Cole, H H, Garrett W N (Eds) 1980 *Animal Agriculture: the Biology Husbandry and Use of Domestic Animals* 2nd ed., W H Freeman, San Francisco

Commission of the European Community 1987 *The State of the Environment in the EEC*, Brussels

Condon, R W 1968 Estimation of Grazing Capacity on Arid Grazing Lands, in: Stewart, G A (Ed.) *Land Evaluation*, Macmillan, Canberra, Australia, 112–24

Cooke, G W 1967 *The Control of Soil Fertility*, Crosby Lockwood

Cooke, G W 1972 *Fertilising for Maximum Yield*, Crosby Lockwood

Cooke, G W, Williams, R J B 1970 Losses of Nitrogen and Phosphorus from Agricultural Land, *Water Treatment and Examination* 19, 253–76

Cooke, G W, Pirie, N W, Bell, G D H 1977 The Management of Inputs for Yet Greater Agricultural Yield and Efficiency, *Philosophical Transactions of the Royal Society* London B281, 73–301

Cooke, G W 1977 Waste of Fertilisers, *Philosophical Transactions of the Royal Society* London B281, 231–41

Cooke, G W 1981 Managing the Nutrient Cycle, *Proceedings of the Nutrient Society* 40, 295–314

Cooper, J P 1970 Potential Production and Energy Conversion in Temperate Arid and Tropical Grasses, *Herbaceous Abstracts* 40(1), 1–15

Cooper, J P (Ed.) 1975 *Photosynthesis and Productivity in Different Environments*, Cambridge University Press

Cooper, J P, Tainton, N M 1968 Light and Temperature Requirements for the Growth of Tropical and Temperate Grasses, *Herbaceous Abstracts* 36(2), 167–76

Countryside Commission 1984 *Agricultural Landscapes: a Second Look*, Cheltenham

Cox, G W, Atkin, M D 1979 *Agricultural Ecology: Analysis of World Food Production Systems*, W H Freeman, San Francisco

Craswell, E T, Vlek, P L G 1979 Fate of Fertiliser Nitrogen Applied to Wetland Rice, in: International Rice Research Institute, 175–92

Crosby, D G 1973 The Fate of Pesticides in the Environment, *Annual Review of Plant Physiology* 24, 467–92

Croxhall, H E, Smith, L P 1984 *The Fight for Food: Factors Limiting Agricultural Production*, Allen and Unwin

Cunningham, E P 1979 Cattle Populations in Relation to their Ecological Environment, in: Bowman, J, Susmel, P (Eds) *The Future of Beef Production in the European Community*, Martinus Nijhoff, The Hague, 153–69

Datta, S K de 1975 Upland Rice Around the World, in: *Major Research in Upland Rice*, International Rice Research Institute, Philippines, 2–11

Datta, S K de 1979 Results from Recent Studies in Nitrogen Fertiliser Efficiency in Wetland Rice, *Outlook on Agriculture* 12(3), 125–34

Datta, S K de 1981 *Principles and Practice of Rice Production*, International Rice Research Institute, Philippines

Davidson, D A 1980 *Soils and Land Use Planning*, Longman

Davies, D B 1975 Field Behaviour of Medium Textured and Silty Soils, in: MAFF Technical Bulletin 29, 52–75

Davies, D B 1981 Soil Degradation and Soil Management in Britain, in: Boelns, D, Davies, D B, Johnston, A E (Eds) *Soil Degradation*, Institute of Land and Water Management Research, Netherlands

Davies, J, Eagle, D J, Finney, J B 1982 *Soil Management* 4th ed., Farming Press

Davies, W, Skidmore, C L 1966 *Tropical Pastures* Faber & Faber

Davis, N E 1972 The Variability of the Onset of Spring in Britain, *Quarterly Journal of the Royal Meteorological Society* 98, 763–77

Dennett, M D 1980 Variability of Annual Wheat Yields in England and Wales, *Agricultural Meteorology* 22, 109–11

Dennett, M D, Elston, J, Speed, C B 1981a Rainfall and Crop Yield in Seasonally Arid West Africa, *Geoforum* 12, 203–9

Dennett, M D Elston, J, Speed, C B 1981b Climate and Cropping Systems in West Africa, *Idem*, 193–202

Dent, D, Young, A 1980 *Soil Survey and Land Evaluation*, Allen and Unwin

Department of Environment 1972 *Soil Capability Classification for Agriculture*, Quebec, Canada

Department of the Environment 1984 Standing Technical Advisory Committee on Water Quality. *Fourth Biennial Report* February 1981–1983, HMSO

Dix, H M 1981 *Environmental Pollution*, J Wiley and Son

Donald, C M, Hamblin, J 1976 The Biological Yield and Harvest Index of Cereals as Agronomic and Plant Breeding Criteria, *Advances in Agronomy* 28, 361–405

Doorenbos, J, Pruitt, W C 1977 Crop Water Requirements, *Irrigation and Drainage Paper 24*, FAO, Rome

Douglas, G K 1984 *Agricultural Sustainability in a Changing World Order*, Westview, Boulder, Colorado

Dowdell, R J 1982 Fate of Nitrogen Applied to Agricultural Crops with Particular Reference to Desertification, *Philosophical Transactions Royal Society* London B296, 263–73

Drosokin, D, Heady, E D 1976 Farming Practices, Environmental Quality and the Energy Crisis, *Agriculture and Environment* 3, 1–13

Duckham, A N 1963 *An Agricultural Synthesis: The Farming Year*, Chatto and Windus

Duckham, A N, Jones, J G W, Roberts, E H (Eds) 1976 *Food Production and Consumption: the efficiency of human food chains and nutrient cycles*, North Holland Pub. Company

Duckham, A N, Masefield, G B 1970 *Farming Systems of the World*, Chatto and Windus

Dutt, G R, Hutchinson, C F, Garduno, M A (Eds) 1981 *Agriculture in Arid and Semi-Arid Regions*, Commonwealth Agricultural Bureau.

Dyson-Hudson, N 1980 Strategic Resource Exploitation among East African Savanna Pastoralists, in: Harris, D R, 1980, 171–84

Eadie, J 1984 *Trends in Agricultural Land Use: The Hills and Uplands*, ITE Symposium No. 13, Monkswood

Eadie, J, Cunningham, I 1971 Efficiency of Hill Sheep Production Systems, in: Wareing, P F, Cooper, J P, 239–49

Eagle, D J 1971 *Residual Value of Applied Nutrients*, MAFF Technical Bulletin 20, HMSO

Eagle, D J 1975 ADAS Ley Fertility Experiments, in: MAFF Technical Bulletin 29, 344–59

Eden, M J 1978 Ecology and Land Development: the Case of the Amazonian Rainforest, *Transactions of the Institute of British Geographers* 3(4), 444–63

Ennis, W B, Jr, Dowler, W M, Klassen, W 1975 Crop Protection to Increase Food Supplies, *Science* 188, 593–8

Evans, L T 1975 (Ed.) *Crop Physiology: some case histories*, Cambridge University Press

Evans, L T 1976 The Two Agricultures: Renewable and Resourceful, *Journal of Australian Institute of Agricultural Science* 42, 222–3

Evans, L T, Wardlaw, I F 1976 Aspects of Comparative Physiology of Grain Yield in Cereals, *Advances in Agronomy* 28, 301–59

Evans, L T 1980 The Natural History of Crop Yields, *American Science* 68, 388–97

Evans, R 1980 Characteristics of Water Eroded Fields in Lowland England, in: De Brodt, M, Gabriels, D (Eds) *Assessment of Erosion*, 77–87

Evans, R 1983 Accumulated Water Erosion in Soils of England and Wales, in: Prendergast, A G (Ed.) *Soil Erosion*, Commission of the EEC, Brussels, 27–87

Evans, S A, Hough, M 1984 Effects of Several Husbandry Factors on the Yield of Winter Barley at Four Sites for Each of Four Years, *Journal of Agricultural Science* 10, 555–60

Evans, S A, Neild, J R A 1981 The Achievement of Very High Yields of Potatoes in The United Kingdom, *Journal of Agricultural Science* 97, 391–6

FAO 1973 *Irrigation, Drainage and Salinity: an International Source Book*, Hutchinson

FAO 1976 *A Framework for Land Evaluation* Soils Bulletin 32, Rome

FAO 1974 *Shifting Cultivation and Soil Conservation in Africa*, Soil Bulletin 24, Rome

FAO 1978 *Agro-Ecological Zones*, Report on the Project Vol. 1 *Methodology and Results for Africa*, Rome

FAO 1980 *Land Evaluation Criteria for Irrigation*, World Soil Report No. 5, Rome

FAO 1982 *FAO Production Yearbook 1982*, Rome

FAO 1982 *Weeds in Tropical Crops*, Rome

Farmer, B H 1981 The Green Revolution in South Asia, *Geography* 66, 202–7

Fiddian, W E H 1973 The Changing Pattern of Cereal Growing, *Annals Applied Biology* 75, 123–49

Fienhel, H, Muir, A 1984 *Soil Salinity*, Van Nostrand

Fletcher, W W 1974 *The Pest War*, Blackwell

Floate, M J S 1970 Mineralisation of Nitrogen and Phosphorus from Organic Materials of Plant and Animal Origins and its Significance in the Nutrient Cycle of Grazed Upland and Hill Soils, *Journal of British Grassland Society* 25, 295–302

Forbes, T J, Dibb, C, Green, J O, Hopkins, A, Pred, S 1980 *Factors Affecting the Productivity of Permanent Grassland: a National Farm Survey*, Grassland Research Institute, Hurley, Maidenhead

Fota, H D, Shafer, J W 1980 *Soil Geography and Land Use*, Wiley and Sons, New York

Francis, C A 1986 *Multiple Cropping*, Macmillan

Frissel, M J (Ed.) 1978 *Cycling of Mineral Nutrients in Agricultural Ecosystems*, Development in Agriculture and Managed Forest Ecology 3, Elsevier, Amsterdam

Fuller, W H 1979 Management of Saline Soils, *Outlook on Agriculture* 10(1), 13–20

Gair, R, Jenkins, J E E, Lester, E 1972 *Cereals, Pests and Disease*, Farming Press

Gibbon, D, Holliday, R, Mattei, F, Luppi, G 1970 Crop Production Potential and Energy Conversion Efficiency in Different Environments *Experimental Agriculture* 6, 197–204

Gibbon, D, Pain, A 1985 *Crops of the Drier Regions of the Tropics*, Longman

Gibson, T E (Ed.) 1978 Weather and Parasitic Animal Disease, Technical Note 159, *World Meteorological Organisation*, Geneva

Gilg A W 1975 Development Control and Agricultural Land Quality, *Town and Country Planning* 43(9), 387–9

Gimingham C H 1975 *An Introduction to Heathland Ecology*, Oliver & Boyd

Gimingham C H 1976 *Ecology of Heathlands* Chapman & Hall

Gleissman, S R, Garcia, E R, Amador, A M 1981 The Ecological Basis for The Application of Traditional Agricultural Technology in The Management of Tropical Agro-Ecosystems, *Agro-Ecosystems* 7, 173–85

Gloyne, R W 1972 Long Range Weather Forecasts and The Farmer, *Scottish Agriculture* 51, 261–7

Goodland, Robert J A, Watson, Catherine & Ledec, George 1984 *Environmental Management in Tropical Agriculture*, A Westview Replica ed., Boulder, Colorado

Gourou, P 1980 *The Tropical World*, Longman

Grasser, J K R 1982 Agricultural Productivity and The Nitrogen Cycle, *Philosophical Transactions Royal Society* London B296, 303–14

Greenhalgh, J F D 1976 The dilemma of animal feeds and nutrition, *Animal Feed Science and Technology* 1, 1–7

Greenland, D J 1977 Soil Damage by Intensive Arable Cultivation: Temporary or Permanent? *Philosophical Transactions Royal Society* London B281, 193–208

Greenland, D J 1981 Soil Management and Soil Degradation, *Journal Soil Science* 32, 301–22

Greenland, D J 1985 Upland Rice, *Outlook on Agriculture* 14, 21–8

Greenland, D J, Lal R (Eds) 1977 *Soil Conservation and Management in The Humid Tropics*, J Wiley and Son, New York

Grennfelt, P (Ed.) 1984 *The Evaluation and Assessment of the Effects of Photochemical Oxidants on Human Health, Agricultural Crops, Forestry Material and Visibility*, Swedish Environmental Research Institute, Goteborg

Grieve, I C 1980 The Magnitude and Significance of Soil Structure Stability Declines Under Cereal Cropping, *Catena* 7, 79–85

Grigg, D 1974 *Agricultural Systems of The World: an Evolutionary Approach*, Cambridge University Press

Grist, 1975 *Rice*, 5th ed., Longman

Grove, A T (1978) Geographical Introduction to the Sahel, *Geographical Journal* 144, 407–15

Gudmundsson, O (Ed.) 1980 *Grazing Research at Northern Latitudes*, Agricultural Research Institute Kildna, Vol. 108 NATO Series A, Life Science

Gupta, V S 1975 *Physiological Aspects of Dryland Farming*, Oxford University Press

Hacker, J B (Ed.) 1982 *Nutritional Limits to Animal Production from Pastures*, Commonwealth Agricultural Bureau

Hafez, E S E 1968 *Adaptation of Domestic Animals* Lea & Febiger, Philadelphia

Hagan, R M, Halse, R R, Edminster, T W (Eds) 1967 *Irrigation of Agricultural Lands*, American Society Agronomy, Madison, Wisconsin

Haines, M 1982 *An Introduction to Farming Systems*, Longman

Hall, A E, Cannell, G H, Lawton, H W 1979 *Agriculture in Semi-Arid Environments*, Ecological Studies 34, Springer-Verlag, Berlin

Hall, H T B 1985 *Diseases and Parasites of Livestock in the Tropics* 2nd ed., Longman

Hamblin, A P, Tennant, D 1981 The Influence of Tillage on Soil Water Behaviour, *Soil Science* 132, 233–9

Hamblin, A P 1985 The Influence of Soil Structure in Water Movement, Crop Root Growth and Water Uptake, *Advances in Agronomy* 38, 95–152

Hanna, L W 1983 Agricultural Meteorology, *Progress in Physical Geography* 7(1), 329–44

Harlan, J R 1976a Plant and Animal Distribution in Relation to Domestication, *Philosophical Transactions Royal Society* London B278, 13–25

Harlan, J R 1976b The Plants and Animals that Nourish Man, in: *Food and Agriculture*, Scientific American Book, W H Freeman, San Francisco, 55–68

Harris, D R 1961 The Distribution and Ancestry of the Domestic Goat, *Proceedings Linnean Society* 173, 79–91

Harris, D R 1972 The Origin of Agriculture in the Tropics, *American Scientist* 60(2), 180–93

Harris, D R 1976 The Ecology of Swidden Cultivation in the Upper Orinoco Rain Forest, Venezuela, *Geographical Review* 614, 75–95

Harris, D R 1980 *Human Ecology in the Savanna Environment*, Academic Press

Harrod, M F 1975 Field Behaviour in Light Soils, in: MAFF Technical Bulletin 29, 22–51

Harwood R R, Price E C 1976 Multiple Cropping in Tropical Asia, in: American Society of Agronomy, 11–40

Heady, H F 1975 *Rangeland Management*, McGraw Hill, New York

Heady, H F, Heady, E B 1982 *Range and Wildlife Management in the Tropics*, Longman

Healy, M J, Ibery, B W (Eds) 1985 *The Industrialisation of The Countryside*, Geography Books

Heath, S B, Roberts, E H 1981 The Determination of Potential Crop Productivity, in: C R W Spedding (Ed.) *Vegetable Productivity*, MacMillan, 17–49

Heathcote, R L 1983 *The Arid Lands: Their Use and Abuse*, Longman

Helbaek, M 1950 Botanical study of the contents of Tollund Man, *Aarbøger for Nordisk Oldkyndighed og Historie* 325–41

Hill, D S, Waller, J M 1982 *Pests and Diseases of Tropical Crops: Principles and Methods of Control* (2 vols), Longman

Hill, T A 1977 *The Biology of Weeds*, Institute of Biological Studies No. 79, Edward Arnold

Hillel, D (Ed.) 1972 *Optimising the Soil Physical Environment Towards Greater Crop Yield*, Academic Press

Hillel, D 1987 *Efficient Use of Water in Irrigation*, World Bank Technical Paper No. 64, Washington DC

Hills, E S 1966 *Arid Lands – A Geographical Appraisal*, Methuen

Hogg, W H 1971 Weather Forecasting for Agriculture, *Agriculture* (London) 78, 352–5

Holliday, R H 1976 The Efficiency of Solar Energy Conversion by the Whole Crop, in: Duckham *et al*, 127–46

Holmes, W 1973 Tillage crops, in: *Organic Resources of Scotland*, Ed. Joy Tivy, Oliver & Boyd, 141–63

Holmes, W 1977 Choosing Between Animals, *Philosophical Transactions Royal Society* London B281, 121–37

Holmes, W 1980 Ed. *Grass: Its Production and Utilisation*, CAB International

Holzen, W, Numata, N. (Eds) 1982 *Biology and Ecology of Weeds*, Dr W Junk, The Hague

Hood, A E M 1976 Nitrogen, grassland and water quality, *Outlook on Agriculture* 8, 320–7

Hood, A E M 1982 Fertiliser Trends in Relation to Biological Productivity within the United Kingdom, *Philosophical Transactions Royal Society* London B296, 315–28

Hopkins, L D 1977 Methods for Generating Land Suitability Maps: a Comparative Evaluation, *Journal American Institute Planning* 43(4), 386

House of Lords Select Committee on Science and Technology 1984, *Agriculture and Environmental Research*, HMSO

House of Lords Select Committee on The European Community 1984, *Agriculture and the Environment*, HMSO

Hudson, J P 1977 Plants and Weather, in: Landsberg, J J and Cutting, C V, 187–201

Hunt, I V 1973 The Grass Crop, in: *The Organic Resources of Scotland*, Joy Tivy (Ed.) Oliver & Boyd, 122–38

Hurst, G W, Smith, L P 1967 Grass Growing Days, in: Taylor, J A (Ed.) *Weather and Agriculture*, Pergamon Press, 147–55

Hutchinson, T C, Havers, M 1980 *The Effect of Acid Precipitation on Terrestrial Ecosystems*, Plenum Publishing Corp., New York

International Rice Research Institute 1978 *Soils and Rice*, Los Banos, Philippines

International Rice Research Institute 1979 *Nitrogen and Rice*, Los Banos, Philippines

Institute of Terrestrial Ecology 1984 *Agriculture and the Environment*, National Environmental Research Council, HMSO

Isaac, E 1970 *Geography of Domestication*, Prentice Hall, New Jersey

Issar, A 1980 The Reclamation of a Desert by the Consideration of Ancient and Modern Water Systems, *Outlook on Agriculture* 10(8), 393–6

Jacks G V 1954 *The Soil*, Nelson

Jacks, G V, White, R O 1939 *The Rape of the Earth: a World Survey of Soil Erosion*, Thomas Nelson

Jackson, I J 1977 *Climate, Water and Agriculture in the Tropics*, Longman

Janick, J, Schery, R W, Words, F W, Rutten, V W 1981 *Plant Science: an Introduction to World Crops* 3rd ed., W H Freeman, San Francisco

Janzen, D H 1973 Tropical Agro-ecosystems, *Science* 182, 121–9

Jeans, D N 1977 *Australia – a Geography*, Angus & Robertson, Sydney

Jenkin, D 1980 *Agriculture and the Environment*, Proceedings NERC/ITE, Monkswood Experimental Station, Symposium No. 13

Jennings, P R 1974 Rice Breeding and the World Food Production, *Science* 185, 1085–8

Jennings, P R 1976 The Amplification of Agricultural Production, *Scientific American* 235(3), 180–94

Jewell, J N C 1975 *Energy, Agriculture and Waste Management*, Ann Arbor Science

Jewell, F A 1980 Herbivore ecology in African Savannas, in: Harris, D R, 353–82

Jewitt, T N, Law, R D, Virgo, K J 1979 Vertisol Soils of the tropics and the sub-tropics: their management and use, *Outlook on Agriculture* 10(1), 33–40

Jones, M J 1973 The Organic Matter Content of Savanna Soils of West Africa, *Journal of Soil Science* 24, 42–53

Jordan, A M 1980 Trypanosomiasis control and land use in Africa, *Outlook on Agriculture* 10, 123–9

Joss, P J, Lynch, P W, Williams, O B (Eds) 1987 *Rangelands: a Resource under Siege*, Proceedings Second International Rangeland Conference, Cambridge University Press

Juo, A S R, Lal, R 1977 The Effect of Fallow and Continuous Cultivation on the Chemical and Physical Properties in Alfisols, West Africa, *Plant and Soil* 4, 567–84

Keay, R W J 1978 Temperate and Tropical Agriculture Contrasts, in: Hawkes J G (Ed.) *Conservation and Agriculture*, Duckworth, 243–8

Kipps, M S 1970 *Production of Field Crops*, 6th ed, McGraw-Hill, New York

Kirby, M, Morgan, R P C 1980 *Soil Erosion*, John Wiley and Son

Klingauf, P 1981 'Inter-relationships between pests and climatic factors', in: Bach, W *et al*, 285–302

Klingebiel, A A, Montgomery, P H 1961 *Land Capability Classification*, Department Agricultural Soil Conservation Service Handbook No. 210, US Dept. of Agriculture, Washington DC

Kogan, F N 1983 Perspectives on Grain Production in the USSR, *Agricultural Meteorology* 28, 213–17

Korda, V A 1977 *Arid Land Irrigation in Developing Countries*, Pergamon Press

Kowal, J M, Kassam, A H 1978 *Agricultural Ecology of Savanna: A Study of West Africa*, Clarendon Press

Kranz, J, Schmitterer, H, Kuch, K 1978 *Diseases Pests and Weeds in Tropical Crops*, John Wiley and Son

Lal, R 1984 Soil Erosion from Tropical Arable Lands and its Control, *Advances in Agronomy* 37, 183–240

Lal, R, Greenland, D J (Eds) 1979 *Soil Physical Properties and Crop Production in the Tropics*, John Wiley and Son

Landsberg, H E (Ed.) 1981 *World Survey of Climatology*, Elsevier, Amsterdam

Landsberg, S S, Cutting, C V (Eds) 1977 *Environmental Effects on Crop Physiology*, Academic Press

Langer, R H, Hill, G D 1982 *Agricultural Plants*, Cambridge University Press

Lazenby, A, Downes, E M 1982 Realising the Potential of British Grasslands: Some Problems and Potentials, *Applied Geography* 2, 171–88

Lee, J 1977 Land Valuation Should be Based on Productivity, *Farm and Food Research* 8, 187–91

Le Houeron, H N (Ed.) 1980 *Browse in Africa*, ILCA, Addas Ababa

Leohr, R C & Hart, Samuel, A 1970 Changing Practices in Agriculture and their Effect on the Environment, *CRC Critical Review Environmental Control* 1, 69–99

Lewis, W M, Phillips, J A 1976 Double cropping in eastern United States, in: American Society of Agronomy, 41–50

Liang T, Khan, M A, Manrique, L A, Parma, R 1986 Rating Land for Crop Introduction, *Agricultural Systems* 21, 107–27

Lie, T A, Mulden, E G (Eds) 1971 Biological Nitrogen Fixation in Natural and Agricultural Habitats, *Plant and Soil* (special volume), 231–64

Litomo, W H, Dexter, A R 1981(a) Tilth Mellowing, *Journal Soil Science* 32, 187–201

Litomo, W H, Dexter, A R 1981(b) Soil Friability, *Journal Soil Science* 32, 203–13

Livingstone, B E 1916 A single index to represent both moisture and temperature conditions, *Physiological Research* 1, 421–40

Lockeretz, W 1977 *Agriculture and Energy*, Academic Press

Lockhart, J A R, Wiseman, A J J 1983 *Introduction to Crop Husbandry*, 5th ed., Pergamon Press

Loomis, R S, Gerakis, P A 1975 Productivity of Agricultural Ecosystems, in: Cooper J P (Ed.), 593–621

Loomis, R S, Williams, W A, 1963 Maximum Crop Productivity – an Estimate, *Crop Science* 3(1), 67–72

Loomis, R S, Williams, W A, Hall, E A 1971 Agricultural Productivity *Annual Review of Plant Physiology* 22, 431–68

Lourbin, L G 1981 Continuous Cultivation and Soil Productivity in the Semi-arid Savanna: The Influence of Crop Rotation, *Agronomy Journal* 73, 317–63

Low, A J 1972 The Effect of Cultivation on Structure and Other Physical Characteristics of Grassland and Arable Soils, 1945–1970 *Journal Soil Science* 23, 363–80

Low, A J 1975 Ley fertility experiment at Jeallots Hill, in: MAFF Technical Bulletin 29, 360–87

Lowe, P 1981 *Countryside Conflicts*, Gower, Aldershot

Lynch, J M 1983 *Soil Biotechnology: Microbiological Factors in Crop Productivity*, Blackwell Press

McCormack, D E, Stocking, M A 1986 Soil Potential Ratings: an Alternative Form of Land Evaluation, *Soil Survey and Land Evaluation* 6, 37–41

MacDonald, R B & Hall, F G 1980 Global Crop Forecasting, *Science* 208, 670–8

MacDonald, L H (Ed.) 1982 *Agroforestry in the African Humid Tropics*, Tokyo University Press

McDowall, R E 1974 The Environment Versus Man and his Animals, in: Cole, H H, Ronning, M (Eds) *Animal Agriculture*, Freeman, San Francisco, 455–69

Mabbut, J A, Floret, C 1980 Case Studies in Desertification, *Natural Resources Research* 18, UNESCO, Paris

Mabbut, J A 1984 A New Global Assessment of the Status and Trends of Desertification, *Environmental Conservation* 11(2), 106

Mabbut, J A 1985 Desertification of the Worlds Rangelands, *Desertification Control Bulletin* 12, 1–11

Marbut, C F 1935 *Soils of the United States* in *American Atlas of Agriculture*, UDSA, Washington DC

Mahadevan, P 1968 The Relations Between Climatic Factors and Animal Production, in: *Agrometeorological Methods, Book 7*, UNESCO, Paris, 115–25

Marshall, T J 1972 Efficient Management of Water in Agriculture, in: Hillel, D (Ed.), 11–22

Mason I L 1984 *Evolution of Domestic Animals* Longman

Mather, J R 1974 *Climatology: fundamentals and applications*, McGraw-Hill

Meiri, A, Shalhvert, J 1973 Crop Growth under Saline Conditions, in: Yaron *et al*, 277–90

Mellanby, K 1970 *Pesticides and Pollution*, Collins

Mellanby, K 1981 *Farming and Wildlife*, Collins

Michael, A M 1978 *Irrigation: Theory and Practice*, Vikas Publishing Press, New Delhi

Milthorpe, F L, Moorby, J 1979 *An Introduction to Crop Physiology*, Cambridge University Press

Ministry of Agriculture, Food and Fisheries 1970 *Modern Farming and the Soil*, HMSO

Ministry of Agriculture, Food and Fisheries 1974 *Land Capability Classification*, Technical Bulletin 30, HMSO

Ministry of Agriculture, Food and Fisheries 1975 *Soil Physical Conditions and Crop Production*, Technical Bulletin 29, HMSO

Ministry of Agriculture, Food and Fisheries 1976a *The Agricultural Climate of England and Wales*, Technical Bulletin 35, HMSO

Ministry of Agriculture, Food and Fisheries 1976b *Climate and Drainage*, Reference Book 434, HMSO

Ministry of Agriculture, Food and Fisheries 1976c *Organic Manures*, Bulletin 210, HMSO

Ministry of Agriculture, Food and Fisheries 1976d *Agriculture and Water Quality* Technical Bulletin 32, HMSO

Ministry of Agriculture, Food and Fisheries 1979 *Agriculture and Pollution*, 7th Report Royal Commission on Agricultural Pollution, Cmnd. 7644, HMSO

Ministry of Agriculture, Food and Fisheries 1980 *Inorganic Pollution and Agriculture*, Reference Book 326, HMSO

Ministry of Agriculture, Food and Fisheries 1982 *Irrigation*, Bulletin No. 138, 4th ed., HMSO

Ministry of Agriculture, Food and Fisheries 1983 *To Plough or Not to Plough* Primer 1 HMSO

Monod, T (Ed.) 1975 *Pastoralism in Tropical Africa*, (International African Institute), Oxford University Press

Monteith, J L 1966 Physical Limitations to Crop Growth, *Agricultural Progress* 40, 9–23

Monteith, J L 1972a Solar Radiation and Productivity in Tropical Ecosystems, *Journal Applied Ecology* 9, 747–60

Monteith, J L 1972b Weather and the Growth of Crops, Amos Memorial Lectures, *Malling Research Institute Station Report for 1971*, 21–34

Monteith, J L 1977 Climate and the Efficiency of Crop Production in Britain, *Philosophical Transactions Royal Society* London B281, 277–94

Monteith, J L 1981a Does Light Limit Crop Production, in: Johnson C G (Ed.) *Physiological Processes Limiting Plant Productivity*, Butterworth, 23–38

Monteith, J L 1981b Climatic Variation and the Growth of Crops, *Quarterly Journal Royal Meteorological Society* 107(454), 760–74

Monteith, J L, Elston, J F 1971 Microclimatology and Crop Production, in: Wareing and Cooper, 23–42

Moore, N W (Ed.) 1966 Pesticides in the Environment and their Effect on Wildlife, *Journal Applied Ecology* Vol. 3, Supplement

Moore, N W 1977 Agriculture and Nature Conservation, *Bulletin British Ecological Society* 8, 2–4

Moormann, F R, van Breeman, N 1978 *Rice, Soil, Water, Land*, IRRI, Los Banos, Philippines

Moormann, F R, van Wambeke, A 1978 The Soils of the Lowland Rainy Tropical Climates: Their Inherent Limitations for Food Production and Related Climatic Constraints, Plenary Papers Vol. 2, *11th International Congress Soil Science*, 272–91

Morgan, R A 1986 Changes in the Breeding Avi Fauna of Agricultural Land in Lowland Britain, *Proceedings 18th International Ornithological Congress*, 588–93

Morgan, R P C 1980 *Soil Erosion*, Longman

Morgan, R P C (Ed.) 1981 *Soil Conservation Problems and Perspectives*, John Wiley

Morgan, W B 1980 *Agriculture in the Third World: A Spatial Analysis*, Bell Hyman

Morris, R M, Potts, G R 1976 *Grasses and Cereal Ecosystems*, The Open University Press

Morris, Dick 1977 *Systems Behaviour. Module 6. The Structure and Management of Ecosystems*, The Open University Press

Mosteck, A, Walsh, J E 1981 Corn Yield Variability: Weather Patterns in the USA, *Agricultural Meteorology* 25, 111–24

Munton, R J C 1983 Agriculture and Conservation: What Room for Compromise, in: Warren, A, Goldsmith, F B *Conservation in Perspective*, John Wiley, 353–72

Murata, Y, Matsushima, S 1975 Rice, in: Evans, 73–100

Murdoch, J C 1980 The Conservation of Grass, in: Holmes, W (Ed.), 174–216

Murdoch, W W 1975 Diversity, Complexity, Stability and Pest Control, *Journal of Applied Ecology* 12, 795–807

Muttsson, L 1985 *Forestry and Reindeer Herding*, Geographical Report 8, Department of Geography, University of Lund, Sweden

National Academy Science 1974 *Productive Agriculture and a Quality Environment*, Washington DC

Nature Conservancy Council 1977 *Nature Conservation and Agriculture*, HMSO

Newman, J, Wang, J Y 1959 Defining Agricultural Seasons in Middle Latitudes, *Agronomy Journal* 51, 579–82

Nicholson, M J 1985 The Water Requirements of Livestock in Africa, *Outlook on Agriculture* 14(4), 156–64

Nienwolt, S 1975 *Tropical Climatology: An Introduction to the Climates of the Low Latitudes*, John Wiley

Nightingale, H I 1974 Soil and Ground Water Salinisation Beneath Diversified Irrigation Agriculture, *Soil Science* 118, 365–73

Nir, D 1974 *The Semi-Arid World: Man on the Fringe of the Desert*, Longman

Norman, M J T, Pearson, C J, Searle, P G E 1984 *The Ecology of Tropical Food Crops*, Cambridge University Press

Northcliff, S 1983 Soil Cultivation, *Progress in Physical Geography* 7(2), 247–55

Northcliff, S 1987 Developments in Soil and Land Evaluation, *Progress in Physical Geography* 11(2), 283–91

Norton, B W 1982 Differences between species in forage quality, in: *Nutritional Limits to*

Animal Production from Pastures, Ed. J B Hacker Commonwealth Agricultural Bureau, London, 89–110

Nye, P H, Greenland, D J 1964 Changing the Soil After Clearing Tropical Forest, *Plant and Soil* 21(1), 101–12

O'Connor, R J, Shrubb, M 1986 *Farming and Birds*, Cambridge University Press

O'Connor, R J, Shrubb, M 1986 Recent Changes in Bird Populations in Relation to Farming Practices in England and Wales, *Journal Royal Agricultural Society England* 147, 132–44

Oelhaf, R C 1978 *Organic Agriculture*, John Wiley, New York

Oelsligle, R E 1976 Soil Fertility Management in Tropical Multiple Cropping, in: American Society of Agronomy, Special Publication No. 27, 275–92

Oguntoyinbo, J S 1981 Climatic Variability and Food Crop Production in West Africa, *Geojournal* 5, 139–49

O'Hare, Greg 1988 *Soils, Vegetation, Ecosystems*, Oliver & Boyd

Okigbo B N and Greenland D J 1976 Inter-cropping systems in Tropical Africa, in: *Multiple Cropping*, American Society of Agronomy, Special Publication No. 27, 63–102

Olson, G W 1982 *Soils and the Environment*, Chapman and Hall, New York

O'Riordan, T 1982 *Putting Trust in the Countryside*, Nature Conservancy Council, HMSO

Osborne, D D, Schneebergen 1983 *Modern Agricultural Management*, Reston, New York

Osbourn, D F 1980 The Feeding Value of Grass and Grass Products, in: Holmes, W (Ed.), 69–124

Page, C C 1972 Arable Crop Rotation, *Journal Royal Agricultural Society England* 133, 98–105

Palmer, W C 1968 Keeping track of crop moisture conditions nationwide: the new crop moisture index *Weatherwise* 21, 156–61

Patterson, G T, MacIntosh, E E 1976 Relationship Between Soil Capability Class and Economic Returns from Grain Corn Production in S W Ontario *Canadian Journal Soil Science* 56, 167–76

Pearson, C J (Ed.) 1984 *Control of Crop Productivity*, Academic Press

Peel, L, Tribe, D E 1983 *Domestication and Use of Animal Resources*, World Animal Science, Elsevier, Amsterdam

Peel, S 1985 Efficiency of Temperate Grasslands: Lessons from the United Kingdom and New Zealand, *Outlook on Agriculture* 15(1), 15–20

Penman, H L 1948 Natural Evaporation from Open Water, Bare Soil, and Grass, *Proc. Roy. Soc. Lond.* Ser. A 193, 120–46

Penman, H L 1949 The Dependence of Transpiration on Weather and Soil Conditions, *Canadian Journal Soil Science* 1, 74–89

Penman, H L 1956 Evaporation: an introductory survey, *Netherlands Journal of Agricultural Science* 4, 9–29

Pereira, H C Agricultural Science and the Traditions of Tillage, *Outlook on Agriculture* 6, 211–12

Pereira, C, Hamblin, M J, Mansell-Mouldin, M 1987 Scientific Aspects of Irrigation, *Philosophical Transactions Royal Society* London 317, 193–737

Perfect, T J 1986 Irrigation as a factor influencing the management of agricultural pests, *Philosophical Transactions Royal Society* London A316, 347–54

Perrin, R M 1977 Pest Management in Multiple Cropping Systems, *Agro-Ecosystems* 3, 93–118

Pfander, W H 1971 Animal Nutrition in the Tropics – Problems and Solutions, *Journal Animal Science* 33, 843–4

Phillips, E D 1961 World Distribution of the Main Types of Cattle, *Journal of Heredity* 52(1), 207–13

Pimentel, D, Dritschilo, W, Krummel, J, Kutzman, J 1975 Energy and Land Constraints in Food Protein Production, *Science* 190, 754–61

Pimentel, D, Terhune, E C, Dyson-Hudson, R, Rocherau, S, Samis, R, Smith, E, Denman, D, Reifschneider, Shepard, M 1976 Land Degradation: Effects on Food and Energy Resources, *Science* 194, 149–55

Pimentel, D (Ed.) 1978 *World Food, Pest Losses and the Environment*, American Association Advancement Science, Washington DC

Pimentel, David and Marcia (Eds) 1979 *Food Energy and Society*, Arnold

Pimentel, D (Ed.) 1980 *Handbook of Energy Utilisation in Agriculture*, CRC Press Inc, Boca Raton

Pimentel, D, Fast, S, Berardi, G 1983 Energy Efficiency of Farming Systems, *Agriculture Ecosystems and Environment* 9, 359–72

Pimentel, D, Hall, C W 1984 *Food and Energy Resources*, Academic Press

Pollard, E, Hooper, M D and Moore, N W 1974 *Hedges*, Collins

Power, J F 1981 Nitrogen in the Cultivated Ecosystem, in: Clark, F E, Rosswall, T (Eds) *Terrestrial Nitrogen Cycles*, Stockholm SNERC Biological Bulletin 33

Purnell, R E 1980 Tick Borne Diseases as a Barrier to Efficient Land Use *Outlook on Agriculture* 10, 230–4

Pyke, M 1970 *Man and Food*, Weidenfeld & Nicolson

Quispel, A (Ed.) 1974 *The Biology of Nitrogen Fixation*, Frontiers of Biology 33, North-Holland Publishing Co., Amsterdam

Rands, M R W, Sotherton, N W 1986 Pesticide use on Cereal Crops and Changes in the Abundance of Butterflies in Arable Farmland in England *Biological Conservation* 36(1), 71–82

Rao, S N S 1984 *Current Developments in Biological Nitrogen Fixation*, Edward Arnold

Rapp, A 1974 *A Review of Desertification in Africa – Water, Vegetation and Man*, Report No. 1, Secretariat for International Ecology, Stockholm

Rapp, A, Hellden, V 1979 *Research on environmental monitoring methods for land use planning in African drylands*. Lund Universitets Naturgeografiska Institution. Rapporter och Notiser Nr. 42

Raymond, W F 1977 Farm Wastes, *Biologist* 24(2), 10–15

Reece, C H 1985 The Role of the Chemical Industry in Improving the Effectiveness of Agriculture. *Philosophical Transactions Royal Society* London B310, 201–13

Reinig, P 1977 *Desertification: its Causes and Consequences*, United Nations Secretariat, Pergamon Press

Reith, J W S, Inkson, R H E, Caldwell, K S, Simpson, W E, Ross, J A M 1984 Effect of Soil Type and Depth on Crop Production *Journal of Agriculture Science* 103, 377–86

Rhoades, J D 1972 Quality of Water for Irrigation, *Soil Science* 113, 277–84

Richards, P 1985 *Indigenous Agriculture: Ecology and Food Production in West Africa*, Hutchinson Press

Ridley, A O, Hedlin, R A 1980 Crop Yield and Soil Management in the Canadian Prairies Past and Present, *Canadian Journal Soil Science* 60, 393–402

Ritchie, J C, Folle, R F 1983 Conservation Tillage: Where To From Here? *Journal Soil and Water Conservation* 38(3), 267–9

Ritchie, J T 1981 Soil Water Availability, *Plant and Soil* 58, 327–38

Roberts, R A, Darrell, N M, Lane, P 1983 Effects of Gaseous Air Pollutants on Agriculture and Forestry in the United Kingdom, *Advances in Applied Biology* 9

Rowett Research Institute and The Hill Farming Research Organization 1974 *Farming the Red Deer*, HMSO

Rosswall, T (Ed.) 1980 *Farming Systems in West Africa in Relation to Nitrogen Cycling*,

SCOPE/UNEP, International Nitrogen Unit, Royal Swedish Academy of Sciences, Stockholm

Rudeforth, C C 1975 Storing and Processing Data for Soil and Landuse Capability Surveys, *Journal Soil Science* 26, 155–68

Russell, E W 1973 *Soil Conditions and Plant Growth* 10th ed. Longman

Russell, E W 1977 Organic matter and soil fertility, *Philosophical Transactions Royal Society London B*, 97, 215

Ruthenberg, H 1976 *Farming Systems in the Tropics* 2nd ed. Oxford University Press

Sakamoto, C M 1978 The Z-index as a Variable for Crop Yield Estimation, *Agricultural Meteorology* 19, 305–25

Samways, J 1981 *Biological Control of Pests and Weeds*, Institute Biological Studies in Biology No. 132, Edward Arnold

Sanchez, P A, Buol, S W 1975 Soils of the Tropics and the World Food Crisis, *Science* 188, 598–605

Sanchez, P A 1976 *Properties and Management of Soils in the Tropics*, J Wiley, New York

Sanchez, P A, Salinas, J G 1981 Low Input Technology for Managing Oxisols and Ultisols in Tropical America, *Advances in Agronomy* 34, 280–406

Sanchez, P A, Vallachi, J H, Brandy, D E 1983 Soil Fertility Dynamics after Clearing a Tropical Rain Forest in Peru, *Soil Science America Journal* 47, 1171–8

Sandford, S 1983 *Management of Pastoral Development in the Third World*, J Wiley, New York

Sasanon, S 1985 Acid Rain: A Review of the Current Controversy. *Soil Use and Management* 1, 34

Sasanon, S 1985 The Nitrate Issue, *Soil Use and Management* 1, 102

Savant, N K, Datta, S K de 1982 Nitrogen Transformation in Wetland Rice Soils, *Advances in Agronomy* 35, 241–303

Schmit-Nielson, K 1956 *Desert Animals*, Oxford University Press

Schwabe, Calvin 1980 Management and Disease, in: Cole, H H, Garrett, W H, 694–710

Schwanitz F 1966 *The Origin of Cultivated Plants* Cambridge, Harvard University Press (Mass.)

Scottish Agricultural Colleges 1987 *Farm Management Handbook*, publication No. 168, 7th ed., 1986–7

Scrimshaw, N S, Young V R 1976 The Requirements of Human Nutrition, *Scientific American* 235(3), 51–65

Seeman, J, Chirkov, Y I, Lomas, J, Primault, B (Eds) 1979 *Agrometeorology*, Springer-Verlag, Berlin

Shainberg, I, Shalhevet, J (Eds) 1984 *Soil Salinity Under Irrigation: Processes and Management*, Springer-Verlag, Berlin

Sharma, R C, Smith, E L 1986 Harvest index in winter wheat, *Crop Science* 26, 1147–52

Shaxton, T F 1981 Determining Erosion Hazard and Land Use Capability: a Rapid Substractive Method, *Soil Survey and Land Evaluation* 1(3), 40–3

Sheail, J B D 1987 *Pesticides and Wildlife: a Case History of Conservation*, Cambridge University Press

Sibma, L 1977 Maximisation of Arable Crop Yields in the Netherlands, *Netherlands Journal Agricultural Science* 25, 278–87

Silvey, V 1978 The Contribution of New Varieties to Increasing Cereal Yield in England and Wales, *Journal Institute Agricultural Botany* 14, 367–84

Simmons, I G 1980 Ecological Functional Approaches to Agriculture in Geographical Contexts, *Geography* 65, 305–16

Simpson, Ken 1980 *Soil* Longman

Simpson, Ken 1986 *Fertilisers and Manures*, Longman

Singh, I D, Stoskopf, N C 1971 Harvest Index in Cereals, *Agronomy Journal*, 63, 224–6

Skidmore, E L, Kumari, M, Larsen, W E 1979 Crop Residue Management for Wind Erosion Control in the Great Plains, *Journal Soil and Water Conservation* 39, 90–4

Slesser, Malcolm 1975 in: Lenihan, J, Fletcher, W, Energy requirements for Agriculture, *Food, Agriculture and Environment*, Blackie, 1–20

Slootmaker, A J 1974 Aims and Objectives in Breeding Cereal Varieties, *Outlook on Agriculture* 3, 133–40

Smilde, K W 1972 The Influence of the Changing Pattern of Agriculture on Fertiliser Use, *Proceedings Fertiliser Society* No. 120

Smith, C V 1970 Weather and Machinery Work Days, *University College Wales Aberystwyth* 17–20

Smith, D F, Hill D M 1975 Natural Agricultural Ecosystems, *Journal Environmental Quality* 4(2), 143–5

Smith, L P 1970 *Weather and Animal Disease*, Technical Note No. 113, WMO Geneva

Smith, L P 1971a *The Significance of Winter Rainfall over Farmland in England and Wales*, MAFF, Technical Bulletin No. 24, HMSO

Smith, L P 1971b Assessment of the Probable Date-of-Return of Soil to Moisture Capacity in the Autumn, *ADAS Quarterly Review* 2, 71–5

Smith, L P 1975 Modes of Agricultural Meteorology, *Developments in Atmospheric Science* 3, Elsevier, Amsterdam

Smith, L P, Davis, J H R H 1972 Autumn Cereal Sowing and Date-of-Return to Field Capacity in the Soil, *ADAS Quarterly Review* 6, 36–40

Smith, M E 1985 *Agriculture and Nature Conservation in Conflict: the Less Favoured Areas of France and the United Kingdom*, Langholm

Smith, R T, Atkinson, K 1975 *Techniques in Pedology*, Elk Science

Soane, B D (Ed.) 1983 *Compaction by Agricultural Vehicles: a Review*, Scottish Institute Agricultural Engineering, Technical Report No. 5

Soane, B D 1985 Traction and Transport Systems as Related to Cropping Systems, *International Conference Soil Dynamics Proceedings* 5, 818–915

Soane, B D, Stafford, J V 1983 Tillage Developments Around the World, *Span* 26(3), 103–5

Soper, M R H, Carter E S 1985 *Modern Farming and the Countryside*, Association of Agriculture

Spedding, C R W 1969 The Agricultural Ecology of Grassland, *Agricultural Progress* 44, 1–23

Spedding, C R W 1975 *Biology and Agricultural Systems* Academic Press

Spedding, C R W 1976 The Biology of Agriculture, *Biologist* 23, 72–80

Spedding, C R W 1979 *An Introduction to Agricultural Systems*, Applied Science Publishers

Spedding, C R W 1985 (Ed.) *Fream's Agriculture*, 17th ed., Royal Agricultural Society, England

Spedding, C R W, Diekmahns, E C 1972 Grasses and Legumes in British Agriculture, *CAB Bulletin 49*

Spedding, C R W, Walsingham, J M, Hoxey, A M 1981 *Biological Efficiency in Agriculture*, Academic Press

Spencer, J E, Stewart, N R 1973 The Nature of Agricultural Systems, *Annals Association American Geographers* 63(4), 529–44

Spoor, G 1975 Fundamental Aspects of Cultivation, in: MAFF Technical Bulletin 29, 128–49

Sprent, Janet I 1987 *The Ecology of the Nitrogen Cycle*, Cambridge University Press

Squires, V R 1978 Distance Trailed to Water and Livestock Response, *Proceedings 1st International Rangeland Congress* 431–4, Denver

Stanhill, G (Ed.) 1984 *Energy and Agriculture*, Springer-Verlag, Berlin

Stanhill, G 1986 Irrigation in Arid Lands, *Philosophical Transactions Royal Society* London A316, 261–73

Staniforth, A R 1974 Cereal Straw Production and Utilisation in England and Wales, *Outlook on Agriculture* 14, 194–200

Stamp, L D 1947 *The Land of Britain: Its Use and Misuse* 3rd ed. 1962, Longman

Steinhardt, R, Trafford, B D 1974 Some Effects of Subsurface Drainage and Ploughing on the Structure and Compactability of a Clay Soil, *Journal Soil Science* 25, 138–52

Stevenson, F J (Ed.) 1982 *Nitrogen in Agricultural Soils*, Agronomy Monograph 22, American Society Agriculture, Madison, Wisconsin

Stewart, W D P, Rosswall, T (Eds) 1982 The Nitrogen Cycle, *Philosophical Transactions Royal Society London* B296, 299–576

Stobbs, E H 1979 Tillage Practices in the Canadian Prairie, *Outlook on Agriculture* 10(1), 21–6

Storie, R E 1954 Land Classification as used in California for the appraisal of land for taxation purposes *Transactions of the 5th International Congress Soil Science* 3, 407–12

Strickland, A H 1960 Ecological Problems in Crop Pest Control in: Wood, R S K (Ed.) *Biological Problems Arising from the Control of Pests and Diseases* Institute of Biology, 1–6

Strickland, A H 1976 Crop Losses in the Field: Control and Some Consequences, *Journal Science Food Agriculture* 27, 702–3

Sturrock, F G, Calthie, J 1980 *Farm Mechanisation and the Countryside*, Occasional Paper No. 12, School of Geography, University of Cambridge

Sutherland J A 1968 *Introduction to Farming*, Angus Robertson

Swanson, E R, Nyankori, J C 1979 The Influence of Weather and Technology on Corn and Soybean Yield Trends, *Agricultural Meteorology* 20, 327–42

Sys, C, Verhey, W 1977 Land Evaluation for Irrigation of Arid Regions by Use of the Parameter Method, *Transactions 10th International Congress Soil Science* 5, 149–55

Szaboles, I 1979 *Review of Research on Salt Affected Soils*, Natural Resources Research 15, UNESCO, Paris

Talbot, M 1984 Yield Variability of Crop Varieties in the United Kingdom *Journal Agricultural Science* 102(2), 315–23

Tarrant, J R 1987 Variability in World Cereal Yields, *Transactions Institute British Geographers* (new series) 12(3), 315–26

Terman, G L 1979 Volatilisation Losses of Nitrogen as Ammonia from Surface Applied Fertilisers, Organic Amendments and Crop Residues, *Advances in Agronomy* 31, 189–223

Terry, R, Powers, W L, Olson, R V, Murphy, L S, Rubison, R M 1981 The Effect of Feed Lot Runoff on the Nitrate-Nitrogen Content of Shallow Aquifers, *Journal of Environment Quality* 10, 22–6

The Royal Society 1985 *The Nitrogen Cycle of the United Kingdom: A Study Group Report*, Cambridge University Press

The Royal Society 1986 *Technology in the 1990s: Agriculture and Food*, Cambridge University Press

Thomasson, E A (Ed.) 1975 *Soils and Field Drainage*, Soil Survey Technical Monograph No. 7, Harpenden

Thomasson, A J, Youngs, E G 1975 Water Movement in the Soil, in: MAFF Technical Bulletin 29, 218–28

Thompson, L M 1975 Weather Variability, Climatic Change and Grain Production, *Science* 188, 535–41

Thorne, D W 1979 Climate and Crop Production Systems, in: Thorne, D W, Thorne, M D (Eds) *Soil Water and Crop Production*, Avi Publishing Company, 17–29

Thornthwaite, C W 1948 An approach towards a rational classification of climate *Geographical Review* 38(1) 55–94

Tinker, P B H 1979 Uptake and Consumption of Soil Nitrogen in Relation to Agronomic Practices, in: Hewitt, J H, Cutting, C V (Eds) *Nitrogen Assimilation of Plants*, Academic Press, 101–22

Tinker, P B (Ed.) 1981 *Soils and Agriculture*, Vol. 2, Blackwell Science Publications

Tinker, P B, Widdowson, F V 1982 Maximising Wheat Yields and Some Causes of Yield Variation, *Proceedings Fertiliser Society* 211, 149–84

Tiplett, G V, Van Doren, D M 1977 Agriculture Without Tillage, *Scientific American* 236, 28–33

Tivy, Joy, O'Hare, Gregory 1981 *Human Impact on the Ecosystem*, Oliver & Boyd

Tivy, Joy 1981 *Biogeography: the role of plants in the ecosystem*, 2nd ed., Longman

Tivy, Joy 1973 (Ed.) *Organic Resources of Scotland: their use and misuse*, Oliver & Boyd

Tivy, Joy 1987 Nitrogen Cycling in Agro-Ecosystems, *Applied Geography* 7, 93–111

Tomlinson, T E 1971 Nutrient Losses from Agricultural Land, *Outlook on Agriculture* 6(6), 272–8

Trentath, B R 1974 Biomass Productivity of Mixtures, *Advances in Agronomy* 26, 177–210

Tribe, D E *et al* 1970 *Animal Ecology, Animal Husbandry and Wildlife Management*, National Resources Research, UNESCO, Paris

Trudgill, S T, Briggs, D J 1971–81 Soils and Landuse Potentials, *Progress in Physical Geography* 1: 319–21, 2: 321–32, 3: 283–299, 4: 282–95, 5: 274–95

Turner, I I, Brush, B L, Stephen B 1987 *Comparative Farming System*, Guildford Press, New York

Turner, H G 1974 The Tropical Adaptations of Beef Cattle: An Australian Study, *World Animal Review* 13, 16–21

United Nations 1977 *Desertification its Causes and Consequences*, Pergamon Press

UNESCO/FAO 1979 *Tropical Grazing and Land Ecosystems: a State of Knowledge Report*, UNESCO, Paris

Unger, P W, McCallan, T M 1980 Conservation Tillage Systems, *Advances in Agronomy* 33, 2–53

Unger, P W 1982 Management of Crops on Clay Soils in the Tropics, *Tropical Agriculture* 59(2) 110–112

USDA 1980 *Report and Recommendations on Organic Farming* Washington DC

USDA 1981 *Council on Environmental Quality. National Agricultural Land Study, Final Report*, Washington DC

Utomo, W H, Dexter, A R 1981b Tilth Mellowing, *Journal Soil Science* 32(2), 187–202

Utomo, W H, Dexter, A R 1981 Age Hardening of Tropical Soils, *Journal Soil Science* 32(3), 335–50

Valli, V J 1968 Weather and Plant Disease Forecasting, in: *Agrometeorological Methods, Book* 7, UNESCO, Paris, 341–6

Van Emden, H F, Williams, G C 1974 Insect Stability and Diversity in Agro-ecosystems, *Annual Review Entomology* 19, 455–75

Van Heemst, H D J 1985 The Influence of Weed Competition on Crop Yield, *Agricultural Systems* 18, 81–93

Vera, R R, Sere, C, Tergas, L E 1987 Development of improved grazing systems in the Savannas of tropical America, in: Joss *et al*, 107–10

Vink, A P A 1975 *Land Use in Advancing Agriculture*, Springer-Verlag, Berlin

Waggoner, P E 1968 Meteorological data and the agricultural problem, in: *Agrometeorological Methods*, UNESCO, Paris, 25–38

Wang, J Y 1972 *Agricultural Meteorology* 3rd ed. Milieu Information Service, San Jose, California

Ward, G M, Sutherland, T M, Sutherland, J T 1980 Animals as an Energy Source in Third World Agriculture, *Science* 208, 570–1

Ward, R C 1975 *Principles of Hydrology*, 2nd ed., McGraw Hill

Ware, G W 1983 *Pesticides: Theory and Application*, Freeman, San Francisco

Wareing, P F, Allen, E J 1977 Physiological Aspects of Crop Choice, *Philosophical Transactions Royal Society London* B281, 107–19

Wareing, P R, Cooper, J P (Eds) 1971 *Potential Crop Production*, Heinemann

Watambe, I 1979 Use of Symbiotic and Free Living Blue Green Algae in Rice Culture, *Outlook on Agriculture* 13(4), 166–72

Watson, H 1971 Size, Structure and Activity of the Productive System of Crops, in: Wareing, P R, Cooper, J P (Eds), 76–88

Watters, R F 1971 *Shifting Agriculture in Latin America*, FAO, Rome

Webster, A J F 1981 Weather and Infectious Disease in Cattle, *Veterinary Record* 180, 183

Webster, J E, Wilson, P N 1980 *Agriculture in the Tropics*, 2nd ed., Longman

Weiers, C S 1975 Soil Classification and Land Evaluation, *Town and Country Planning* 43, 390–3

Weiers, C S, Reid, I G 1975 *Soil Classification, Land Valuation and Taxation: The German Experience*, Centre European Agricultural Studies Wye College

Weisner, C J 1970 *Climate Irrigation and Agriculture*, Angus and Robertson, Sydney

Westmacott, R, Worthington, T 1974 *New Agricultural Landscapes*, Countryside Commission, Cheltenham

Wetsclaar, R, Ganry, F 1981 Nitrogen Balance in Tropical Agro-ecosystems, in: Dommergues, Y, Diem, H G (Eds) *Microbiology of Tropical Soils*, Martinns Nijhoff, The Hague, 1–36

White, G F 1978 *Environmental Effects of Arid Land Irrigation in Developing Countries*, Technical Note 8, UNESCO, Paris

White, R E, 1987 *Introduction to the Principles and Practice of Soil Science* 2nd ed, Blackwell Scientific Publications, Oxford

Whyte, R C 1976 *Land and Land Appraisal*, W Junk, The Hague

Whyte, R O 1974 *Tropical Grazing Lands: Communities and Constituents*, W Junk, The Hague

Wigglesworth, Vincent B 1965 Biological control of pests *Science Journal* April 40–45

Wilkinson, B 1968 Land Capability – Has it a Place in Agriculture? *Agriculture* 75, 343–7

Wilkinson, B 1975 Field Experience on heavy soils, in: *Soil Physical Conditions and Crop Production*, in: MAFF Technical Bulletin 29, 76–93

Williams, C, Joseph, K T 1970 *Climate Soil and Crop Production in the Humid Tropics*, Oxford University Press

Williams, R J B, Cooke, G W 1981 Some Effects of Farmyard Manure and Grass Residue on Soil Structure, *Soil Science* 92, 30–5

Williams, T E 1980 Hertape Production: Grasses and Leguminous Forage Crops, in: Holmes, W (Ed.), 6–69

Wilson, A D 1974 Water Consumption and Water Turnover of Sheep Grazing Semi-Arid Pasture Communities in New South Wales, *Australian Journal Agricultural Research* 22, 339–47

Wilson, P N 1973 Livestock Physiology and Nutrition, *Philosophical Transactions Royal Society London* B276, 101–12

Wilson, R T 1984 *The Camel*, Longman

Wittmus, H, Olson, L, Lane, O 1975 Energy Requirements for Conventional and Minimum Tillage, *Journal Soil and Water Conservation* 30, 72–5

Wittee, S H 1979 Future Technological Advances in Agriculture and their Impact on the Regulatory Environment, *Bioscience* 29, 603–10

Wood, L J 1980 Energy and Agriculture: Some Geographical Implications, *Tijdschrift Voor Economic en Sociale* 72, 224–34

Worthington, E B (Ed.) 1977 *Arid Land Irrigation in Developing Countries* Pergamon Press

Wrigley, G 1981 *Tropical Agriculture* 4th ed. Longman

Yao, Augustine Y M 1981 2. Agricultural Climatology, in: Landsberg, H R (Ed.) *World Survey of Climatology*, Elsevier, Amsterdam, 189–298

Yaron, E, Danfors, E, Vaadia, Y (Eds) 1973 *Arid Zone Irrigation*, Ecological Studies Vol. 5, Chapman and Hall/Springer-Verlag, New York

Yarwood, C E 1970 Man-Made Plant Disease, *Science* 168, 218–20

Yeates, N T M, Schmitt, P 1974 *Beef Cattle Production*, Butterworth

Yoshida, S 1977 Rice, in: Alvim and Kozlowski, 57–87

Young, A 1974a Some Aspects of Tropical Soils, *Geography* 59(3), 233–9

Young, A 1974b The Nature and Management of the Poorer Tropical Latosols, *Outlook on Agriculture* 10, 27–32

Young, A 1976 *Tropical Soils and Soil Survey*, CUP

Young, A, Goldsmith, P F 1977 Soil Survey and Land Evaluation in Developing Countries, *Geography* 143, 407–38

Young, A, Wright, A C S 1980 *Rest Period Requirements for Tropical and Sub-Tropical Soils Under Annual Crops*, FAO, Rome

Zeuner, F E 1963 *The History of Domesticated Animals*, Hutchinson

Zonn, I 1977 Irrigation of the Worlds Arid Lands, *World Crops and Livestock* 29, 72–3

Index